The Life of Jesus

POETRY BY TOBY OLSON

Maps

Worms into Nails

The Brand

Pig/s Book

Vectors

Fishing

The Wrestlers & Other Poems

Changing Appearance: Poems 1965–1970

Home

The Life of Jesus

by TOBY OLSON

A New Directions Book

"... he is not averse to the joy of pain.
Like all solitary figures, he would rather
be lacerated and touched than avoided. It
is the untouchable who is deprived of every-
day raptures."
    EDWARD DAHLBERG, *Because I Was Flesh*

Parts of this novel first appeared in the following magazines, to whose editors and publishers grateful acknowledgment is made: *Center, Fireweed, For Now, Helicon, Lillabulero, New Directions in Prose and Poetry, The Mysterious Barricades*. The poems "Apocryphal Dream: Incest" and "Apocryphal Dream: The Father" were included in *The Wrestlers & Other Poems* (1974), by Toby Olson, and reprinted by permission of Barlenmir House, Publishers.

The epigraphic quotation is taken from Edward Dalhberg's *Because I Was Flesh* (Copyright © 1963 by Edward Dalhberg), published by New Directions Publishing Corporation.

Manufactured in the United States of America
First published clothbound and as New Directions Paperbook 417 in 1976
Published simultaneously in Canada by McClelland & Stewart, Ltd.

Library of Congress Cataloging in Publication Data

Olson, Toby.
    The life of Jesus.

    (A New Directions Book)
    1. Jesus Christ—Fiction. I. Title.
PZ4.0523Li [PS3565.L84] 813'.5'4 76–8210
ISBN 0–8112–0613–0
ISBN 0–8112–0614–9 pbk.

New Directions Books are published for James Laughlin
by New Directions Publishing Corporation,
333 Sixth Avenue, New York 10014

For my mother

The Life of Jesus

NATIVITY: A PROLOGUE

I think there is one thing we hold in common; that is, at least in our secret hearts, we each believe we were born of a virgin. For whose father is not remembered, sitting in the corners of rooms or raging, sick or always out of the house, or there and in the way? And what mother does not have a separate life of her own in our memories?

Perhaps it is different for women; but for men, fathers are finally ludicrous. Always, slightly symbolic or vague, they stand somewhere between a host and a guest in the household. It is a woman's world.

It is not finally a question of sexual intercourse at issue, for no one really believes anything comes from that, certainly not ourselves: young men of genius, somewhere above the human. But each of us, at the same time, demands some marvelous history for ourselves, and what solves these things is the idea of Immaculate Conception. Remember when our fathers speak of the time before our birth, they say: "Why, in those days you were only a twinkle in my eye." They wink feebly at us. We turn to our mothers, who sit, peaceful and smug, looking beyond the walls of the room, into some authentic family.

When I was ten years old, I was Monarch of the Gypsy Band. *My scepter waved o'er every land,* and when I moved out, among the rocks, to the center of where my people were gathered, *all bent beneath my rule and sway, and all, as King, my rule obeyed.* That was the year, being Monarch, that I remember most about Nativity, the school I attended in the lower grades. And the season I remember best was the Christmas season.

My mother had made fine garments for me. My father was

getting sicker those days, and she had taken her old half-broken Singer out on the back porch, in order not to disturb my father with noise, to sew things up. Out of an old floppy hat of her own she had made my soft crown, stitching the felt with rhinestones. My cape was cut from an old full skirt of hers, and she had glued "sparkles" to a pair of my old tennis shoes. My satin shirt was hers also, and out of a pair of well-worn knickers she'd made my pants. Occasionally there were fittings, and I remember how, at those times, putting the garments on, I did *feel* like a Monarch or a King.

It was chilly, even in California, during the Christmas season, and we'd play handball, furiously, against the side of the school building to keep warm. I remember how I would imagine myself, playing among my friends wearing my Monarch costume, my cape flowing around my body as I raced for the ball, my sparkling hat fixed tight to my head. I of course never spoke of this to my friends.

The pastor of our flock at Nativity was Father Ghinty. Even then he seemed old — or maybe ageless is better — though he always walked briskly. He always seemed brittle and neuter to me, his body guarded from definition under his black robes. His rectory was across the dirt playground from the church, and I remember vividly the day I first saw him carry one of the undefined objects out and across the wide expanse of the playground and into the church. The object was large; it protruded from under his arm; it had an odd shape, seemed not to have been constructed for transport this way. When he got closer to where we were playing, I saw it was a large plaster sheep. He was of course taking it into the church to begin setting up the Nativity scene. When Father Ghinty passed by us we hummed faintly the tune to a seasonal song that he had formerly chastised us for singing. Passing, he glanced sharply across the sheep's head, looking over the tops of his glasses. But we gave no evidence of guilt. The song went like this:

> We three kings of Orientar
> Tried to smoke a rubber cigar.
> The cigar exploded.
> We imploded . . . .

I don't remember the rest, but I do remember the image that the song evoked in me: three mysterious and exotically garbed men, riding on camels, their various gifts visible, and the bulges of huge black cigars protruding from under their garments.

Around the Christmas season the Irish nuns would add some Bible stories to our lessons. The one I remember most vividly was about Joseph and Mary. They were out for a walk together. They passed by a cherry tree, and Mary asked Joseph to get her some of those ripe cherries from the top of the tree. Joseph is angry. He tells her: "Let the one who got you with child get you the cherries." And with that, a long sinuous limb of the tree lowered itself down from the top. It hung suspended, ripe with the best cherries, in front of the Virgin, right at a level with her waist. All she had to do was reach out. She picked off the cherries with long fingers. Joseph learned a lesson.

All I remember of the meeting of Mary and the Angel is that they were alone when they met; it was in the daytime; both the Angel and Mary seemed, by their comfort in each other's presence, to have met before.

The second thing I saw Father Ghinty carrying was a small donkey. He moved quickly, bent slightly against the wind, walked down and away from the porch of the rectory, the donkey tucked under his arm. Later, entering the church in the early morning for choir practice, I saw that the whole group was there, by the side of the altar, gathered under the cut-away thatched roof of the make-believe manger. The women had sewn real garments for the statue of Mary, but the rest of the figures were of plaster of Paris, though painted and giving the semblance, from a distance at least, of life. From wherever I looked they always seemed farther away than they really were; this was because they were only half the size of their natural models, but not really miniature; they were large enough to give off a hint of possible reality. The figure of Joseph stood somewhat to the side of the crib; he seemed more like a tentative watcher than a participant. There were four animals. There were the Three Wise Kings. The animals, though placid, seemed to each have a twinkle in his painted eyes. The kings seemed awe-struck to me. It was rather as if they had brought their gifts out of some vague discomfort; maybe the gifts were brought for the mother; maybe they brought them out of atonement, maybe guilt. The crib was empty of course, and it seemed so very strange to see them all looking into it, almost as though, if they looked long and hard enough, the Baby Jesus would appear there, magically, of his own accord.

It is not incidental that I think of my mother here as immaculate. On the night before Christmas she finished my garments. They fit me perfectly, and early on Christmas morning I wore them under my raincoat, peddling my bicycle to five o'clock Mass at the church. I sang in the choir that year, and had to bring in my costume to be inspected by the nuns. The Operetta was going to be performed on Christmas night. I carried my own real clothing in a bundle, in the basket of my bike.

I remember how, in my enthusiasm, I got to the church early. It was empty. And after I'd hung up my raincoat, I walked in my Gypsy King costume, silently up, to look at the manger and the gathering. The crib was still empty; all was silent, and for some reason I felt impelled to join them. I crept, holding my cape back, over the altar rail, and moved among the plaster animals and humans, touching them, all but the mother, even scratching the slick hard head of the donkey. Then I heard a cough from the sacristy. I was terrified of being caught, for surely what I was doing hinted of blasphemy. So I silently climbed over the stable railing and hid down between the forelegs of the donkey.

I could see Father Ghinty pass quietly over the steps, down from the altar. He came up to the manger; stood in his black robes surveying it. I saw that he carried nothing in his hands, but that there was a very strange smile upon his face. And then I saw him kneel down beside the crib. And while I was watching, my right hand came up and covered my mouth. For Father Ghinty had reached up under the robes of the Virgin, had extracted a small package, wrapped like a mummy in linen. He unwrapped the linen, revealing the figure of a child. He held up the child, which now seemed to glow with a light of its own, in the palms of his hands, and then he placed the child into the crib. And then he stood, put a finger beside his nose, threw

4

back his head and laughed. And then he turned round and left us.

I was hid away in the manger in appropriate clothing. To-night was the night of my debut. My immaculate mother had dressed me. The light entered the stained-glass windows high in the church wall behind me; it struck against the star hanging from the manger's eaves. The light flashed from the star, bathing the crib with various colors of light. I was the incarnation of a Gypsy King! It was Christmas morning. When Jesus was born, I was there.

THE EARLY YEARS

CARPENTRY

The child grew, waxed strong in spirit, was filled with wisdom, and the grace of God was upon him, but he couldn't cut wood straight. Every time he tried, the saw would bend, the lintel wind up crooked upon the post. If there had been a house he'd built with his own hands, at times of the day the sun would have entered through cracks in the doorframe and left through chinks in the wall on the other side. It would have been a house of light, but cold in an open meadow, and no one could have lived comfortably in it.

It was the same with the pegs, doweling inserted in round holes that were always too big or not quite round, and he would have to split the pegs with wedges that couldn't be properly sanded, and he would throw his tools down, swear under his breath, and go sit in the back yard.

Once, his father, who was a patient man, took great pains to include him in the building of a neighbor's stable. The lengths of fencing were cut by the father, and it only remained for him to hold them in place while his father measured and fitted the crosspieces and inserted the pegs at the proper height and angle. But even here the son was awkward, letting the pieces slip when his father measured, and that evening the neighbor accused the father, charging him with the loss of four sheep that had escaped when the fence collapsed under the weight of their bodies scratching against it. From that day on the boy was sent to work in his mother's kitchen.

His mother was a terrible cook. But in her case, as in the son's, it was not lack of natural wisdom and spirit that prevented her accomplishing such simple tasks. It was rather a degree of abstraction (in both) that would not allow their

7

eyes to fix themselves for very long on the things that were before them.

But his father was a patient man, and though his meals often stuck in his throat, he would never say a word of reproach, but simply nod and manage a smile through his closed teeth while he was eating.

The son learned to make things that resembled bread, and often, not looking at his kneading hands, would see his mother under the olive tree in the back yard, with hands over her breasts, looking up through the branches.

> I met God in the morning
> When my day was at its best
> And his presence came like sunrise
> Like a glory to my breast.
>
> All day long the presence lingered
> All day long He stayed with me
> And we sailed in perfect calmness
> O'er a very troubled sea

. . . she would say, while her pies cooked in the oven, and it was less the prayer than the beauty of her posture that intrigued the son, causing him to watch closely her every movement.

Whether or not the father noticed this growing attachment between mother and son is not clear, but one day he returned from his labors bearing a large dog under his arm, and he presented the mongrel to his son. Immediately the boy named the dog "Hound," and the very next morning he fashioned a large cape for the dog, securing it around the dog's neck with a twine cincture. This was a custom among young boys at the time, and a visitor from somewhere else might have remarked at the way dogs raced through fields and along roadways with capes flung out behind them.

Hound slept under the steps in front of the house, and would sometimes whine at night, and it was not long before the father suggested that the boy build the dog a proper little house, so that he might have some protection against the cold.

As before, the son's carpentry left much to be desired, and

8

though he finished the small peaked roofed house in a day, he had not accounted for the degree of pitch of the patch of land on which he built it, and the house leaned at a precarious angle to the left when one looked at it from the front. Chagrined, the boy waited in the kitchen for his father's return and mild disapproval.

But when the father entered the house his face beamed with a great smile, and he said to the boy, "What a wonderful house you have built!" The mother, saying

> So I think I know the secret
> Learned from many a troubled way
> You must seek Him in the morning
> If you want Him through the day

. . . could be heard from the window, and the boy was amazed at the way his father dissembled, and he went with him out to the yard to look at his structure.

And lo, where a crooked house had stood, now stood a straight house, and the dog was in it, looking at them, his cape resting on his haunches.

All that night, and for three nights, the boy lay awake in his bed wondering about the miracle. And each night, whether in a dream or not he could not tell, he saw Hound in three separate places, each above the ground. On the first night he saw him sitting on a branch of a tree looking in at the window; on the second night the dog was on the roof of the doghouse; on the third night he fancied he saw the dog pass over the trees, literally flying in the sky, his cape rippling out behind him.

But it was not until the fourth day, at dinner time, that the fact of the matter was known for sure. Hound was sitting by the boy's side, which he always did at dinner time, being allowed the bounty of scraps that neither the father nor the son thought it was possible to digest. At a moment when his father was busy nodding and smiling at the mother, who was gazing out of the window at the olive tree, the boy felt the dog's muzzle pushing against his hand. When he looked down he observed that the dog's feet had left the floor and that he

9

was floating in the air, his head at a level with the top of the table. Immediately the boy pushed the dog back down onto his feet, and as soon as it appeared gracious to do so, he excused himself from the table, and taking the dog by the collar, he led him out into the back yard.

It was at this time that the boy first called into action his powers of induction. He had observed that the dog could rise above the ground. He reasoned that this might have something to do with the miracle of the doghouse. He chose as hypothesis that the dog had effected the resurrection of the house through his powers of levitation. Then he measured the distance from the roof of the house to its floor and found that the height was considerably more than that of the dog's haunches. Next he placed the dog in the house and, looking him in the eye, he said "Rise up," and immediately the dog rose, till his back was wedged in the inverted V of the gabled roof. He then removed the dog from the house and pushed on the side of the roof, at which action the house slumped into the posture it had possessed when he had first built it. After returning the house to its favored position, he returned to his room and went to bed.

The next morning his father said to him, "Now you have become a carpenter. Come, we will work together." But the son answered him in the following way:

A man once journeyed to Egypt to increase his business, and while on the road he passed another man who was also going there. When the two men passed a crossroad a third man joined them, and because there were many wild beasts along the way, the three decided they would travel together. A fourth man, coming out of the desert, joined them, and when they had traveled no more than a mile, a fifth man approached them saying, "May I not travel with you?"

After a while, two men leading a donkey came upon them. "Where are you going," they asked, and the five answered in unison: "To Egypt." So the two men joined them.

Now the journey was long and arduous, and the travelers passed through many towns. In the town of Falla they stopped at an inn to refresh themselves, and when they had eaten and were ready to resume their journey, four men

approached them, saying, "We have learned that you are going to Egypt and would like to go with you." And the seven
acquiesced, and the four joined them. And after a while they
came to the gates of the City. And they were eleven, and
were a crowd.

The father was puzzled by this answer, but he took as the
gist of it that the son did not fancy himself a carpenter, and
being a patient man he did not impose this occupation upon
his son.

For the rest of the summer the boy spent his time fishing,
and by devising a way in which his dog could help him at this
task he became very successful at it. Often, when he was returning at the end of the day, well laden with fish, people would
stop him and ask, "Where have you been; where are you going?" And he would answer them, saying, "Don't bother me; I
must be about my father's business."

APOCRYPHAL DREAM: INCEST

A young boy decides
he wants to go to bed with his mother,
who has given him
ample indication
that it's all right to do that.

And because the boy's mother
and father sleep in separate rooms, there's
no problem about the boy's going in
which he does one night
but in the morning they can't get separated.

It seems
the mother's vagina
has for some reason begun to expand
and started to pull the boy back inside her,
which creates a problem
because he is 10 years old.

But thank god he isn't a big boy
and the mother is able
to strap him to her body
with two of her husband's belts, tho
most of him is still outside of her
but if he hunches up in a ball
she can cover him with an old maternity dress.

Now the father is quite shocked
when he notices but the mother is very coy
and delicate and says she's been to the doctor that day
and that she's been pregnant for months.
She tells him their son
left just that morning for camp.

Now the marvelous thing
is that the father becomes very attentive again
to the mother after 10 years of ignoring her,
and begins to bring flowers and candy.
And the mother feels delicate and feminine
again, and the boy keeps moving inside her.

And then after 4 months (
of being fed in the bathroom
of close quiet talks with his mother
of hearing the tenderness between his parents)
even his head goes inside
and the father finally
goes away on a business trip and the mother
goes to an out-of-the-way hospital.

When the father comes back the mother
tells him about the miscarriage
but neither of them are too sad
and take great pleasure
in having their son home from camp.

From that day on
the father sleeps in the mother's bedroom,
the son takes
the bedroom his father has vacated
and begins to get interested in girls.

ABLE TO FLY

Hound would help him fish in the following way: he'd take the bobber in his teeth, six feet from the hook, and hover over the water till he saw a fish. Then he'd lower the line slowly, being careful not to disturb the prey. When the fish saw the hook so close, he'd bite it, and the boy would haul him in.

Fishing bored the boy. But Hound, and the delicate and controlled way he could fly through the air, never did. He himself wanted to fly, and he set out to discover the secret.

He tried everything. He jumped off of high rocks, wearing Hound's cape around his neck; he'd run down hills with the cape held like a kite; he even tried sitting on the dog's back, but his weight was too much for Hound, and it wouldn't work.

Then one day he went fishing far from his house, in a place full of olive trees, but with very few stones on the ground. While he was sitting under one of the trees with Hound's cape over his shoulders, he noticed he'd risen up about a foot above the ground. He was of course elated, but when he tried to reproduce the act while in his bedroom late that night, nothing would happen, and he couldn't figure out the variables.

But the next morning while he was walking out to speak to his mother, who stood beneath the olive tree looking up into the branches, he noticed that the closer he got to her the lighter he felt, until he was walking on air just a small fraction off the ground.

It was months later, and only after much study, that he discovered the secret. There were certain rocks, plentiful near his house, that held him down. Olive trees, on the other hand, produced a kind of reverse gravity for him, and near them (providing there weren't any of the rocks around) he was able to fly.

But the rocks did more than hold him down. Near them his mouth felt common, devoid of the Fabulous City, and the skin of his body refused to tingle.

And so he asked his mother to make a cape for him, and with it over his shoulders he set out to map the land. At first he would only fly a few inches off the ground, and he found that by so doing he could visually calculate how much the rocks would pull him and how much the trees could counteract that pull.

He spent two years in mapping and becoming accustomed to signs that warned of the rocks' existence. By the time he was ready to start his public life he had mastered the terrain for a radius of ten miles, and even when on unfamiliar ground he could tell, with some skill, which areas he must avoid.

When he flew he flew very high in the sky, and at that height he looked only like a large bird, one of those that swoops down often in search of prey. He flew high enough so that if the rocks pulled him he could turn in his dive and navigate over some olive trees.

By the time he was ready to start his public life, he had already become an omen among the people, who would say: "Watch for the great bird in the sky; it brings good fortune!" And whenever bad times or events fell upon a family or individual, there would always be someone who would give of the following advice: "Do not dwell in your grief: Look! Up in the sky!"

SCHOOL CHUMS

It was not that he feared the walk in the cold weather, or the wolves he could hear snarling in the woods beside the road, or even the distance, bringing him home after dark, when there were strange shadows in the trees. It was rather his chums, who, jealous of his knowledge, would often strike him on the playground, treating him as an outsider, that made him dislike going to school.

His mother was little help; though she petted him when he cried, her eye would always wander to the window and he would see this happen. So he took to sitting behind his father's shed, on Saturdays, finding solace in being alone, simply unharassed.

And one day, while sitting there — sun and grass in the low meadow before him — he noticed the glass of a broken jar, and when he got up and stood over it he saw his reflection in each piece of it, at least twenty miniatures of himself. But a few of the miniatures were soiled with dirt and not as clear as the others, and so he wet his finger to clean them, and soon found that if he licked them they became even cleaner.

And while he was licking one, he inadvertantly bit into it, and though he tried to spit it out, a small part of it remained in his mouth. It had a strange taste to him. When he moved his mouth, it ground between his teeth. And so he began to actively chew the glass, and when he had swallowed all of it, he took up another piece of it and ate it also. And soon he had eaten the whole jar, and when he had done so he found that his mouth was tingling and that he was aware of his teeth and gums. He could feel each tooth individually, and they seemed to vibrate and move in his mouth. And soon each tooth began to grow larger to him, but each to a different size and shape. And they began to feel like buildings, his gums like foundations, and soon it was as if he had a whole city in his mouth. He became afraid then and went back to the house, but he did not speak until dinner.

While they were eating, his father said to him: "And how

did you spend your day?" And he said to his father: "O, I sat behind the shed and read some schoolbooks." And as he said the words the nouns in his speech became real things. The shed appeared in the corner of the room, and the books came, stacked in a pile beside his chair. But his father was intent on chewing his lamb, and his mother had one eye fixed on the window, and neither of them saw these things happen. The boy immediately drank some wine, and when he spoke again the shed and the books withdrew themselves, and nothing startling happened.

When the boy went to school that Monday, he took a piece of glass in his lunch pail and kept it there till the recess time. And while he was standing alone beside where his chums were playing, he ate the glass and began to speak. He said: "Hound is here beside me," and he felt the dog walking through the streets of the City in his mouth, and then it appeared beside him. But his chums were angry that he should bring his dog to school, and they began to strike him, each taking his turn, until he had blue marks on his body.

The next Saturday he took four large jars behind the shed and broke them all. As he was eating them, feeling the City grow up in his mouth, he began to learn the feel of the glass in his hands. The glass cut him and he began to bleed. And so he spat in his palm to remove the blood, but his mouth was full of ground glass, and he found that if he rubbed it into the cut the blood stopped flowing and his hand felt better. He spat glass into both hands then, and completely coated them with it, and as he did this he felt two dismembered hands clinging to a doorway of one of the buildings in his mouth.

So he got a hammer and every jar he could find, and he broke them all, and chewed them and removed his clothing and began to rub the glass into his body. And when he had finished rubbing the glass into every pore he became aware of himself walking around in his mouth. But the figure he was seemed hollow and without internal substance, so he took a gob of glass mixed with his own spittle and began packing it into his shallow navel. The more glass he rubbed into the opening the more he needed, and soon he began to feel the glass

17

crystals moving like tiny stars through his veins and arteries. It was then that the figure in the City in his mouth became flesh and dwelt in him.

The next Monday, at recess, he stood apart and spoke: "Hound is here beside me," but when his chums began to strike him, their hands became scraped on his body, and they began to hurt and bleed. This made them fear him, but he spoke to them kindly, and they thought him a fine fellow and benevolent, and when he invited them to his house for the following Saturday they all accepted willingly.

On Saturday, before his chums came, he prepared himself, and when they arrived he took them behind the shed, where they sat in a circle around him. He stripped himself, and before their wondering eyes he ate of the glass and rubbed it into his body. Then he began to speak and to call forth buildings out of his mouth. And after he had called forth many buildings and animals and had withdrawn them through negation, he bade them look at the meadow that fell away behind them. And as they were looking, a house of light appeared in the meadow, and it was makeshift, of wood, and light shone through cracks and chinks in the wood. And then he took in his hand a chicken that he had hidden behind a bush, and he took a piece of glass and cut off the chicken's head. And he put the whole of the chicken's neck into his mouth, and while they were watching this, they saw his eyes fill up with blood and become like two suns in his head and they saw his neck convulse as he swallowed the blood. And when he had thrown the chicken aside, he bade them look again at the meadow. And lo, the chinks in the house had closed, and the house now glowed red with a light of its own. And he said to them: "Behold, the temple of my body."

From that day he became respected by his chums, and though they feared him, he took no advantage of them. Neither did he make a spectacle of himself, but from that day remained exceedingly careful how he spoke.

APOCRYPHAL DREAM: THE FATHER

This is a very
traditional
        story about a father
who is cut from tin, and a son
who has surely only
created this image of his father

The mother
plays a very small part in this:
she helps
        the son carry
his father to bed at night;
she reminds the father to
go thru
        motions of eating at dinner time

The father is truly
cut out of tin, and though he has hinges
at his knees and hips
from the waist up
he is very rigid
and his arms are only hinged at the shoulders

The night before
the father takes the son out on weekends
the son creeps into the bedroom
where his father sleeps
and paints
        an appropriate smile
or grin, on the flat surface of the father's face

Now the son imagines
there is a real father
about, that he sleeps under the bed of the tin
father, and that he is rounded
like the son is

19

But the son can't
                    look under there
afraid, that what he imagines
might be true   The mother
plays a very small part in this
tho she tells him always
to sleep in the center of the bed

One day
the father takes the son out in the country
and as long as he stands
with his flat body
cutting the wind
everything goes along fine

But while they are walking
out of a wood and over a rise in a meadow
the son trips
and the father turns to help
him, and is caught by the wind

Now imagine how
the father reaches out to the son
as he is lifted
                    slightly off the ground
he
picks up speed and is waving his legs
and stiff arms
to keep his balance   Think of the son
dumb-struck and helpless
running after
his father who misses the first tree
but is struck
by a hanging limb
and is ripped in half
                    at the abdomen:
the one sound the son hears —
the tearing of a piece of tin

20

Now the mother
plays a very small part in this
and when the son gets home, she says:
where is your father?

   "I suppose he's hiding under the bed
again"   And the mother says
well, be sure to sleep right in the center
of the bed tonight

and before the son goes to sleep
he begins to feel
his skeleton, eating his flesh away
from the inside.

AMAZING GRACE

We could imagine a cup, filled to the brim with good wine. The capacity of the cup increases as the wine is poured. And yet the cup is always filled to the brim, and does not seem to grow in size. We can tell this, because we hold the cup in our hands; yet surely our hands remain the same, and yet the wine keeps pouring. There is something amazing about this.

But when we think of the Virgin Birth, or the Ascension, or Mary up in Heaven, her capacity increasing, full of Grace and growing, there seems nothing amazing about it.

But this is about Jesus — before the Testament is written — who goes off to school. He is respected by his chums. He has done amazing things. There is nothing that should concern him. And yet we find him, locked in his upper room at night, beating his head on the wall — Ann or Nancy or Leah — moaning some name or other, thinking crazy things, putting his cape on; posing in it before the window.

He realized, on his way to school, that he had begun to walk in his own history, and was not surprised to find his chums waiting for him just inside the fence at the edge of the grounds. They cajoled him (though they somewhat feared him), but he remained intractable, thinking somehow — if incorrectly — that to exercise his powers would be to begin to use them up. And so he stood silent among them, though some of them even began to whine, saying: "O, please, bring forth the dog; bring forth some buildings for us." They moved around him and even bowed, but he remained silent.

And even the girls had begun to gather (though somewhat apart from the boys), and as he smiled and looked around and tried to calm them, he saw her looking at him, and it seemed that she smiled in the same way that he himself was smiling, and that she understood, somehow, what was in his mind. "What was that one's name?" he asked one of his chums later. And his chum replied: "Her name is Grace."

Or Ann or Nancy or Leah: it doesn't matter. And yet it matters greatly and quite specifically.

He would watch the deliberate way she placed her pencil on the desk, the incredible cleanliness of her ankles, the placement of her hair, her posture, her elegant beauty of form, her amazing grace. And he took to subtle manipulations of the crowd of chums that always gathered around him; he would move them slowly across the grounds to where she sat with her friends at lunch time, would raise his voice so slightly, and speak with great seriousness and precision. But to speak to her directly, that surely would have been amazing.

And there was no one to talk to, even in veiled terms: his mother looked up at the sky; his father was an impossibility.

And so he is locked in his room; he is beating his head on the wall: Ann or Nancy or Leah, moaning her name: Amazing Grace; the cup is full to the brim. And he calls incredible structures out of his mouth. The room is packed with them; he can hardly move. And yet the vessel is full, and ready to burst!

Adolescent Love. Amazing Grace. This is the Vessel of God!

The things he does to himself, behind his father's shed; even now we cannot speak of them. They are less than amazing.

HIS FATHER'S HOUSE — his own voice

I remember, in winter, rising before the first light: how cold it was! It was my job to get the horses ready. I'd dress and go out to the stable and take down the harnesses that hung like huge intricate belts along the stable wall. I was not a big boy, and the harnesses were always frozen stiff, and I'd have to struggle to crack the hinges loose, before I put the bridles on the horses. And then I'd yoke the horses to the skid. It was so cold. I used to wrap my hands in old cloth, and even with that by the time I was near finished I'd be holding my hands under the horse's steaming nostrils to warm them.

But then my father would come out, and he'd bring with him a huge stone vessel of hot tea, and we would stand there, in the middle of the stable, both holding the wide vessel, warming our hands. My father seldom smiled and almost never laughed, but I remember on those early winter mornings he always had a bright twinkle in his eyes, and I always felt at those times that my father was proud to have me working with him.

The boards were stacked behind the shed, and sometimes we would have to crack the top ones loose with a hammer, because the snow had turned to ice during the cold night and they would be frozen together. We'd pull them apart with a ripping sound, but the ones underneath always seemed to have a certain warmth to my hands, and it was good to lift them, my father holding the other end, and stack them on the skids. I remember feeling that, alone with my father, lifting those boards on those cold, winter mornings, my mother seemed somehow in another life, only playing a very small part in the one between us.

The trees had been cut by men my father hired, who always seemed mysterious to me, because I was not allowed to enter the forest when the heavy work was going on, and the men never entered the house. I would see them often, on bleak snowy days, moving among the trees, and often would hear them yelling in warning; and then a tree would fall, stirring a dense cloud of snow when it struck the ground. By the time I was old enough to have helped in the felling, my father had retired from that aspect of business, and contented himself with the building of small structures that required the "fine work" that he was capable of.

Animals loved my father. And I remember how I'd sit on the lumber while he drove the horses from the yard. He would never touch the reins, but would speak quietly to them, and he would reach out and stroke their haunches when the going was particularly tough or we had gotten bogged down in a snow-bank. "Come now, come now" he'd almost whisper to them, and the horses would pull harder and strain their muscles for him.

Our house was small and sat on a rise in a meadow, a good two hundred yards from the forest. My father had built it of large beams, and where the beams were joined (and it was im-possible to seal them with wood) he'd filled up the chinks with dried moss.

There were five rooms in the house: downstairs, a kitchen, my father's bedroom, my mother's bedroom, and a large family meeting room. Upstairs was only my small bedroom, under the eaves. I remember the family room well. My father had made all of the furniture in it, and had himself made the hearth out of stone and mortar. In the evenings my father would sit in one of the rough chairs and smoke his pipe.

My father worked hard, and consequently would retire early, and most of my memories center around having to be quiet in the evenings while he was sleeping. I seldom saw him in the afternoons, when he was working, building various structures for people, far and wide.

Sometimes my father would come home so exhausted that he was barely able to go through the motions of eating at dinner time. He would fall asleep in his chair, and my mother and I would almost have to carry him to his bedroom. I remember, I'd creep into his room; he slept with his face to the wall, and I would see his heavy breathing in his back, and be reassured that his extreme exhaustion was no affliction, but the condition of his life.

But there is one evening that I remember with a particular warmth (not unmixed with a certain sadness). I was almost twelve at the time, and knew I was soon to be leaving. My other father had spoken to me in dreams, and I had discovered my City. Though I never spoke of these things to him, my father seemed aware of the change in me. I remember sitting in the yard in the early afternoon and being dumb-struck to see my father approaching in his wagon. It was the only time in my recollection that my father returned home before dusk, and added to the amazement was the fact that my father was smiling, and that he bade me help him unload a small lamb from the back of the wagon. Even my mother came out to greet him. And he had brought her flowers and candy.

We spent the rest of the afternoon killing and dressing the lamb, and by the time the sun fell one of its ample legs was turning on a spit over the fire under the hand of my mother. Then my father said, "Let me show you something that we used to do when I was a child." And he bade me fetch a pail of hot water and take it outside. He then put a ladder up to the side of the house, against the icy eaves, and he bade me climb the ladder with a bucket. "Now pour the water over the ice, but mind, do it slowly." And I did so, and was thrilled to see how the water froze as it fell, forming long, delicate icicles, over three feet in length. After we had broken off three of the icicles, my father then produced a bag of colored substance from his pocket (it was sugar), and, insisting that my mother join us, each of us sprinkled the sugar on our icicle, and for a few glorious minutes the three of us stood in the yard, in a circle, munching and nodding at each other. I remember that even my mother, having dropped a glob of melting ice on her garment, could not contain her laughter, and how my father and I pummeled each other lightly on the shoulders.

After we had eaten, my father took out his pipe as usual, but to my amazement he offered me a pipe as well. "Now, what shall we smoke?" he said. "Well, why not try the house!" And with that he got up, and with his fingers he pulled some moss out from between the beams, and ground it to a rough tobacco in his palms. I could hardly contain my laughter, but when he had packed and handed me a pipe, I took part in his mock seriousness, and sat like him, my legs crossed, facing the fire, smoking. The pipe tasted bitter and good.

Such days as that have passed now. And this is the image I have of my father. And whether or not I have created it, I cannot tell. So much has passed, and I have not seen my father for many years. Still, it is memories like these that precede me. I am now in the service of my other father, but the two things must remain separate. In my City there are many fine buildings, and there the country has a beauty never seen in my own real past. But I have prepared a small place, with a house in it that is an exact replica of my own. It is not unoften that I walk there, alone, recalling such memories, from what I have put away.

THE CITY IN HIS MOUTH

The City in his mouth is futuristic, a realized eschatology. A new world, it rises from his gums, which are its pavements, seamless ribbons of sidewalk, that bank at corners, so that a man walking straight can make whatever turn he so desires.

The buildings, in his City, are made of concrete; offices, but mostly temples, without icons, wherein he is worshiped.

Nobody's poor, in his City, but nobody's rich either.

No sun, but the light of his own glass body, the veins visible in it, lights the City.

When he goes on the streets, of his City, he walks in his cape, Hound by his side, a chicken tucked under his left arm, a pocket of glass. When he passes people, he gives lessons on the chicken, or uses it as illustration.

But there is country too! full of olive trees, but no hard stones, where he sometimes walks and talks to his mother, who listens to what he is saying. His father has a fine shop, on the outskirts of his City, and he sometimes helps him build fences, which are not square but of intricate pattern that he identifies, flying over them, at a perfect level with the ground.

But mostly he walks through his City with his chums, who wear the letter 'J' on arm bands and are known to the people, who bow when they pass and touch the hem of his robe. And he bids them to rise, but doesn't touch them, as not to cut them, but directs his chums to touch their eyes or spit on their tongues.

There is no new building, in his City, and seldom reconstruction, but there is this story:

Once a great crack appeared in a building that came as a searing pain in the side of his mouth, and though he bathed it and put a sling around his cheek, it persisted and caused him to turn and writhe and not be able to sleep. And soon his head engineer came to him, and he told the engineer, "Prepare the crack by drilling it smooth." And this was done, and then he called his father to him, and he put fifty stone masons under his father's command. And after his father had built a brace out of wood, the crack was filled with cement. And then the brace was tightened, and the cement hardened, and under his father's direction the masons sanded the area smooth. Thus it is told, of the way his father helped him, in restoring his his City to its perfection.

AT THE GATE — his own voice

It was time to go, and I, having packed my things, walked on
the path that led down hill from the house. I was leaving;
Hound was going before me. He'd romp ahead, bite on a
stone, wiggle, and throw it into the air, and then he would stop,
looking back at me, to see that I was following.

We'd eaten dinner — my father had come home early this
day — in the late afternoon. My mother had put up a small
bundle for me, of food; my father had tied it onto a stick. I carried the stick, with the bundle hanging and swaying, over my
shoulder. I'd kick at a stone, and Hound would run to fetch it;
his cape moved from the motion, slightly, over his haunches.
There was no real wind.

Eating, we didn't talk much. My father said a few things,
about what he'd done that day, but the words seemed to stick
in our throats. My father had talked things over with me, a few
days previous, out in the shed at the back of the house. My
mother had little to say about my leaving; she knew it was time
to happen; I think she knew she'd see me again, and often.

And as I was going, slowly, kicking at stones, I turned back
and saw them. They were standing together, close, at the gate,
in the first hint of the failing light. I was two hundred yards
away. Standing together, their arms at their sides, I could see
my father's hand begin to move.

I don't want to go, I said to myself; and then I said it out
loud: "I don't want to go."

And suddenly, I was back in the house again, standing before the window, looking at them from the back. They were
both naked, though I could see the edge of the clothing that
covered the front of their bodies. My father's buttocks were
heavy, but his hips were narrow; my mother had a slim waist;
her spine stood out, like a rope down the middle of her back.
My father's heels were calloused; his arm had moved halfway
around my mother's body; it was not touching her, but it
looked to be heading for her shoulder. I felt Hound's nose nuz-
zle into my palm. Then came the lines, like a sharp aura, that

30

began to appear around their bodies; and color began to take on shape; the folds of fabric; my mother's green robes came into focus; my father became covered in brown fabric. I kicked the wall below the window: I must leave, I said to myself; and immediately the wall turned to a stone: skittering along the ground; Hound was chasing it. I was out on the path again.

This time, as I looked back, their bodies were naked in front, and I knew I should not look at them. But I could not turn away; my mother had taken a posture halfway tilted at the neck, her head close to my father's shoulder; her arm too had moved, forward, and up to the level of her waist. They stood at the gate; my father's belly was hard and firm, my mother's was soft. I could see dark places, like hair, in their crotches; my mother's breasts were small. And then the clothes began again; they seemed to be formed in the way a child might fill in a figure in a coloring book, after he'd strongly outlined the form.

How can I leave them like this, I thought, there at the gate, touching each other; what will they do in the evenings without me. My cape hung loose over one shoulder. Hound whimpered and nipped at my heels, hoping I'd throw him a stone to fetch. My bundle bobbed at the end of my stick. They stood at the gate, watching.

And then I was moving, backward, slowly; and Hound was moving too. And I could see, over my shoulder, to where they were standing, their clothing beginning to dissolve again from their bodies. I started to call forth fence posts out of my mouth; I called forth stanchions and rooted trees; but when I grabbed for them, in order to hold myself back, they seemed always to be out of reach, and I could only brush them with my hands and pass them. I was getting closer; Hound floated beside me; their clothes were melting away. And then I tried calling out an entire fence to block my way; but instead of the fence, the door of the house flew open; my father's heavy chair moved through it, floating in the air, coming to meet me. And when it reached me, it hit me behind the knees. And then I sat in the chair, Hound held tight in my arms, and kept drifting back, toward them.

They stood at the gate — beside it; and the gate was open,

31

and as I passed them, the clothes on the front of their bodies began to form again, and the backs of their bodies became naked again, as I passed them. The chair entered the open door of the house, and it came to rest, gently, back in front of the window again.

It was then I wept and asked my father to help me. The hand of the father who stood at the gate was hovering over my mother's shoulder. My mother's head was almost touching the father's body. Her own arm was completely extended now, before her; the palm had turned toward her own face.

I wept, knowing I had to go; Hound stood waiting beside me. I called out to my real father: "Help me," I said.

And then I was back on the path again; and then I was kicking at stones, and Hound was chasing them, throwing them into the air, scampering back and forth. And when I reached a place where the path turned, I looked back at the house again.

They were standing together, my father's hand on my mother's shoulder, hugging her to him; my mother's head pressed into my father's neck; she had covered her mouth with her hand.

This is the image fixed in my memory, as I am going away. And it stays with me, and comes as a light in my mind often: my mother, my father, watching me go away. They stand in the failing light, close together, at the gate.

PUBLIC LIFE

FIRST MIRACLE

Already he walks in a strange way. He puts his toe down first, like a man testing the unknown quality of his bath or beginning to enter his foot into a pool of water in a dark forest. And not all his toes at once, in a block: his thigh contracts and pulls up; his leg swings out from the knee; the ankle appears to dislocate; his toes aim at the earth two feet before his body. And just before his foot touches the ground, the toes move singularly; they literally grasp at the earth when his foot touches, conforming to each pebble and clod of dirt. He feels it all through his thin sandal.

He is walking between his parents, who are the same size; he is already slightly taller than they are; his shoulders ride at an even level; the heads of his parents bob, up and down, as they take each step. He does not look at the ground, but fixes his eyes on objects over a hundred yards in the distance. His parents walk in a varying gait down the path; they often talk, back and forth, in front of his body. He is slightly bent at the waist, his neck slightly extended. His cape hides his hanging arms from view; he is thin; his cape seems to breathe in and out as he moves, like the rising/falling wings of a large bird at rest. His shoulders remain on the same horizontal plane; his legs bend, then lengthen as the pitch of the ground changes, keeping him fixed at the same place, moving through air.

His father walks on his right side; his mother walks on his left. His cape occasionally brushes their shoulders when it billows out. They seem to be walking very close together, but the cape breathes out a good eight inches; their bodies are never touching. His father kicks at occasional clods of dirt on the path; his toes sometimes brush against plants. His mother's

34

sandals are studded with woven leather; the weaving catches in vines as she passes; sometimes her sandals rip out plants by their roots.

"Who do you think will be there?"

"Family and neighbors," he answers.

"And people from town? He deals with people from town in his business," she says.

"I don't know," he says. "We didn't talk about that."

They emerge from a wood and enter a broad meadow; the path widens. Across the meadow is another wood. The parents are able to move a few inches away from the son; they still walk abreast, but the path is wider now. His eyes scan the wood, a mile away in the distance; they come to rest, pause, then move on, over the various trees. Each tree is distinct to his vision; he sees them as if he were standing much closer. They slowly transverse the distance, but his view of the various particular growths doesn't change. He can see the eaves of a house through the trees' branches; the house sits in a clearing in the forest. He sees light glint off a window; he dissolves the light. Two aunts and a grandfather sit in the room beyond; he can hear their words distinctly. He extracts his vision, back to the edge of the meadow. He goes through a list of the various kinds of trees. His toes gather around the pebbles and clods of dirt. From behind, his cape falls down straight from his shoulders; only the motion of his legs and feet are visible. The soles of his sandals can be seen as his feet toe down; his heels press last into the path. His head seems to rest on his shoulders, but really his neck is extended forward. The conventional walk of his parents frames him. They occasionally bend toward each other across his body. The mother is asking questions. He identifies trees in the distance. He begins to foresee the possible future.

When we set a table we do it in a uniform way; we place knife and spoon on the right; the fork goes on the left. And above the fork, to the left of the plate itself, we place the salad plate. We do various other things. In a way the setting of a table seems gratuitous, sometimes even absurd. We say that Victoria was left-handed, explaining the salad plate, possibly even the fork. We cut meat with our right hand. But in some cultures we pass the fork in our hands after cutting, and this is a wasted motion; we need to reach cross our plates for our salad, and what if our sleeves are wide? And yet we want to say that manners are imperative, that those who ignore them are inconsiderate, and therefore, that good manners are part of our moral behavior. This is true.

The steward of a landowner was invited to his master's house for drinks in the evening. The time was set for seven. And because the master was a stern man, the steward constantly feared for his job. He vowed to arrive on time, and he directed his wife to dress in her finest garments, and to have his own best clothing laid out for him when he returned from his work at five. And his wife did this. And as they dressed they talked to each other, instructed each other, each from his or her own limited knowledge, as to how they should behave in the master's house. And when the time came to depart for the master's house they found it was raining. But they wrapped old clothes around their shoes, and when they reached the house, they removed the rags, having kept their shoes clean. It was exactly seven o'clock, and they knocked at the door of the master's house.

For a while no one came to open the door, but then the master himself came and opened it. He was dressed in his bathrobe and he was angry. "Why have you come at seven?" he said to his steward. But the steward was disconcerted and could not answer. But the steward's wife answered, saying: "We were invited for drinks at seven." And the master was furious that the steward's wife had spoken for her husband, and he spoke to the steward, saying: "Since you are not even the master of your own house, of your wife, how can I trust you to manage my affairs? Henceforth you have no place in my employ. Go, and see to your wife's behavior." And the steward and the steward's wife left, wondering where they would live and feeling that they had done wrong.

But it was the master who had done wrong to the steward, in failing to tell him the significance of seven o'clock.

In the possible future someone will ask him to do a thing he is unprepared for; nothing difficult, but a thing he will have vague questions about: will it be right do to this thing? What will right mean in the context of the favor? What will doing this thing accomplish? Will he need to reveal himself? They are coming closer to the place where the forest begins, and now he is fixing his gaze on the house that sits in the clearing beyond. There are more trees surrounding the house at a distance. The house is large. Two families are approaching the house from another direction, walking, also along a path, from another meadow, entering the forest from the right. They will reach the house before he and his parents reach it. His parents walk closer to his body now as they enter the forest; the path narrows; his cape is stirred by their bodies brushing against it again. His mother no longer questions his father. They both look ahead now, anticipating the appearance of the house. His father walks with the same shuffling gait he had used all along the way. His mother adjusts her clothing as they go; she gathers her veil in her hands over her breasts; she glances down at her sandals. He sees the front door of the house open; a man walks out of the door and faces the path to the right of them; he waves. The two families that are approaching wave back at him. Then they converge in a group near the door. He walks with his parents along the narrowing path; the path takes a slight turn to the left, then broadens; the trees have been cut down here; the house is now visible through the trees.

The bride reclines in her parents' bedroom, which is situated at the rear of the house, its windows overlooking a garden, the first trees of the forest behind it. The groom opens the door of the bedroom slightly; he looks in at his new wife. The bride's eyes are closed; he thinks she is napping; he closes the door quietly and returns to his preparations. The bride reclines in a half-sitting posture. There are pillows propping her up. Her gown has been spread from her waist, over her parent's bed, like a huge beautiful white fan; it almost covers the bed. Her hands rest together in her lap; her arms are dressed in sleeves of crocheted fabric; the sleeves flare out into intricate cuffs at her wrists; she still wears her veil; the top of the bodice of her dress opens like a flower, then rises up to her neck; her shoes have been placed together beside the bed. Her feet, in white stockings, are crossed at the ankles, and hang over the bed's edge. She is not sleeping; she is merely resting her eyes. Now she is looking out the window. The flowers in the garden wave in a slight breeze, their scent coming in a mixture to her nostrils, which flare slightly as she breathes. She can hear the murmur of voices in another room, the voices of her two aunts and her grandfather. The grandfather lives in the house with her parents; she imagines he is entertaining her aunts with stories. The aunts have arrived early. Her parents expected this. She will go out to greet them when the reception begins. They do not expect her before then. There is a place in the garden the size of a small hoop laid on the ground; this place is empty of flowers. In that section of the garden roses are growing. Two years before, her father had planted, in the now empty circle, her bridal bouquet. The roses now stand in a vase in the room. The empty place in the garden is dark with new turned soil; her father had planted new seed this morning. The selling of flowers is his business.

The bride doesn't sweat in her beautiful dress. The air is perfect in the room, and her dress is adjusted with great care against any wrinkling. She is perfectly comfortable in this position, reclining against the firm pillows; she is not nervous or anxious, and the sounds of the people talking, the men and the women preparing food in the kitchen, the brief appearance of her husband at the door: all these are part of the pleasure of her wedding day. She is perfectly comfortable in time, which passes slowly between each moment, moving with grace. She closes her eyes, and naps.

The father of the bride stands, sad and marvelous, in the open yard. The holy family emerges from the woods a hundred yards to his right. He is greeting the two families; the women embrace him and then stand awkwardly back, while the men shake his hands; one man stands with an arm over the father's shoulder; the other stands before him and grins, shaking his hand, holding onto his upper arm as he does so. The father has never smiled so broadly, and his face seems slightly strange, the corners of his mouth stretched to a virgin extension; he seems awkward, graceless, slightly mechanical in his speech and gesture. This is what the families expect. They greet the father of the bride. He ushers them into his house; his wife stands in the doorway, and then he turns to the path on his right.

The son walks in the space between his parents. His father waves at the smiling host. The mother smiles broadly but does not wave yet; she lowers her arms to her sides, turns her palms out, elevates her hands slightly. The son looks intently at the father of the bride; he is standing before the house now, fifty yards away. He had seen the man walk back out from the door, after he had guided the earlier guests into the house. He had walked out just seven and a half feet, had stopped then, turned, and walked in an arc just at that distance, always keeping his back to the house, the distance remaining always seven and a half feet. He has never seen a man do this before. They are twenty-five yards away now, and the father of the bride seems drawn to come and meet them; he leans forward slightly. The son thinks he could come a little distance away, transverse the shortening space between them, meet them there; but he seems unable to get farther away from the door, and there is no one else coming down the other path. But with all this, he is gracious, is honestly happy to see them; the son can see this, where he waits for them, within the invisible arc. The son searches the bride's father's face for an answer.

And now they have entered the imaginary line. His father is shaking the hand of the host. The son stands awkward between his parents, feels much taller than he really is. His father is bobbing up and down as he shakes the man's hand; he is smiling quite naturally; he embraces the host with familiar ease. The men know each other because of their related businesses, and though the words are new and different, they talk in the

39

same way they would talk were they working together. His mother embraces the father of the bride. The son sees her looking, over the host's shoulder, in at the door, seeing who is there. The son now faces the host on his own; his arm appears from under his cape; it extends to its full length from the shoulder. The host takes the extended hand, shakes it, smiles up at the son's face, releases the hand, puts his own hand flat on the son's shoulder, ushers the holy family into his house.

The wine steward readies the wine. The wedding is over. The bride rests in a room, somewhere in the house. Though most of the wine comes from the town, some of it has been brought from a great distance, over land. He stands in the kitchen, where the casks are lined up against the wall; they are large and full to the brim with wine. Each cask contains the finest quality, appropriate for the various stages of the meal. He lifts the cover of each cask, looks into the moist rich depth of each pool. The groom enters the kitchen; he walks to various places in the room quickly; he looks at the plates of covered food; he even glances into the cask that the wine steward is currently studying; then he walks quickly, out by another door. The wine steward shakes his head and smiles: "The Groom," he thinks to himself; he replaces the wooden cover of the cask and moves on to the next one.

They are going to ask him to do something, and he is going to have to do it. There is really no problem of execution in the matter; the specific is easily distillable from the general request: a brief limited exercise of the will, and no question of ethics involved. But there will be — and this comes to him only in some vague way — other considerations involved, ones that appear to contain matters of the placement of silver on a table or the way one talks to another, from a posture dictated by the forms of specific occasions.

He sits in a chair by a window in the large living room of the house. The room is full up with people. His legs are crossed at the knees; he uncrosses them, finds it difficult to organize the placement of his hands. His father rests very close to him, is talking with men; he sits easily in his robes; he rocks slightly in soft laughter; the men who are talking with him are quite animated in gesture; his father takes part in the conversation with ease: he says his own share of things, but mostly he listens, with grace, to the others. His mother stands in another part of the room, talking with women. He notices a pattern; her eyes begin to move while she is talking; they take in another, ongoing conversation at a distance from where she is standing. She extracts herself from the conversation; she moves, after short periods of time, to talk with other people.

This is his first time in public; or at least, this is his first time sitting in a room full of people who are not all listening to what he might be saying. Actually, he is saying nothing. People sit around him talking, and he nods his head and occasionally gives an impression of joining in talk with them. When he met with the elders he was much younger.

Each person in a public gathering always thinks of himself as the center of what is happening. That is, he proceeds as subject, he observes, he knows what he does and says changes what he perceives as real. He may be only a distant relative or a last-minute guest, but when he picks up an olive from a dish on the table, when he takes his eyes, for a moment, from the central person of honor, the serving lady he looks at, her hands clearing away dishes, becomes for the moment, graced by the central fact of his presence.

Such things are possibly matters of illusion; they are at least arguable, but in this case the place where he sits is the true center. His vision rests on the casks of wine; he sees them through the wall of the room; they rest in a row on the other side. His mother stops talking to a small group of women; she follows her eyes into the kitchen. This is really his first time in public; he is, in a way, tentative in his understanding of the forms. He begins to observe the structure of the movement of the women with food-laden trays, passing, among the various clusters of guests. While he is doing this, he is trying to find pattern to the seemingly erratic movements of the groom. And while he is doing this he is sipping from a glass of wine in his hand. When he allows himself he can taste the faint bite of resin at the back of his throat. But the wine is of the most excellent quality; he is sure no one else can taste the resin; it is very far back; when he closes the buds far back on his tongue the resin fades away, and the wine sparkles, delicious, spilling over his palate. This is the most excellent wine he has ever tasted.

The bride stands in her beautiful gown gracing the room's area. The gown falls in its folds, perfect, from where it is tucked, in at the waist; opens, full as a bell or inverted flower, flows down to her ankles and shoes. This is the only time she will wear this gown, and the gown seems perishable; so finely made, its perfection a part of its preparation. Her body will never be exactly the same as it is today; the last fitting of the gown took place in the early morning; it is a one-day gown, a part of occasion. The bride, the day, the gown: all ripe and perishable, like perfect fruit. And the way she tilts her body in the gown; the way her arms gesture; no one will ever talk to her in the same way again. She sees her mother tending to the bringing of food. Her father sits talking with men. Her husband moves among clusters of people; he enters and exits the room often; on occasion — she can measure it almost exactly in time — he comes up to her, takes her arm at the elbow, brushes her veil with his lips, talks to the people around her. Now she studies the son of her father's colleague, the carpenter. She remembers him from school, knows he's been away from home. She is pleased by the way he fits in; he talks to the guests freely; none of the strange veiled things she had heard about him seem evident. Her father walks rather briskly out of the kitchen; he speaks in the ear of the steward. The two walk back to the kitchen. The carpenter's wife follows them. The bride watches her husband adjust a window, away in the room's far corner; he turns and begins talking with friends. The wife of the carpenter comes out of

42

the kitchen and walks over to her son. She sits beside him and begins to talk. A perfect breeze keeps the room cool. The front door is open. There are no troublesome bugs in the air. Some of the guests stand in the yard talking. The bride smiles and talks with the guests standing around her.

His mother is sitting before him, talking intently into the side of his face. And though he is listening, at the same time he is watching the way the bride moves in her gown. She is beautiful; almost, to him, she is a perishable object of art. She commands her place in the room's area. Never has she looked quite like this. He begins to understand the movements of the groom; he returns at regular intervals to the bride's side. He comes like a moth to a flame. (The son realizes that no one has ever thought of a thing in that particular way before.) His eyes see into the dress to her body. He understands the way the dress is pleated, the way it is attached with stays to her bodice, the intricate way the sewing of the threads has formed the dress. The garments under the dress are made with the same care. The bride has a lovely body; her hair is put up naturally and simply under her veiled crown. He places the flowers in the vase in her parent's bedroom in her arms for a moment, real in his mind. The flowers are the products of careful, almost exquisite, preparations of soil. The petals are symmetrical and perfect. The flowers will close in the room in the evening. When they open next morning they will have begun, already, to die. "It's not my time yet," he says to his mother, who leaves him and goes back into the kitchen. In a few moments his mother returns. This time he acquiesces.

The steward stands, flat-footed, his head bent down, looking into the empty casks. The father of the bride stands, beside him, fidgeting slightly with his clothing. "Are they needing more wine right now?" the steward asks. "Not quite yet," says the father. "But very soon." The son enters the room with his mother and moves to the cask on the far left. He lifts off the wooden cover, reaches in with the dipper; he withdraws it, full to the brim with wine.

The wine is served, he discovers, as a kind of signal that the party is beginning to draw to a close. He sees the various guests sipping at glasses of the wine. Some guests are getting their coats; they return, bid a final farewell to the bride and the groom, who now stands always beside his wife, taking and releasing her elbow, brushing his lips against her veil. When the guests go to the door, the two sets of parents accompany them, bid them farewell as they go. He sees that the steward stands at the kitchen door; men occasionally come up and ask him something; he knows they are asking him a question because of the way the steward responds to their talk with a shrug of his shoulders. He sees now that the bride is watching the steward also; she holds a glass of the wine in her hands. She sips from it; a slight frown appears on her face. Two men are talking behind her. He sees that she is hearing what they are saying, but she does not turn to face them. For a moment she is disconcerted, but then she recovers; nobody notices this but him. The bride doesn't really recover completely. The party ends for her before the guests have gone. He can see that the whole day changes its coloring for her. He sees that the flowers are already beginning to close in the vase in her parent's bedroom. He reads in her mind that she is thinking of the flowers also. With the coming of evening a few bugs appear; they move in the sun's rays beyond the open front door. His father returns to him from a place in the room with his mother. It is time to go.

The three figures walk close together in the beginning of the failing light. They pass through a wood and enter the edge of a meadow. The path widens, but the mother walks still close to

44

the son. The father seems to walk somewhat alone, though only a few inches more from the son's shoulder. He is looking into the trees, watching the day-birds entering, closing their feathers for night. Occasionally the mother talks across the son's body to the father, who seldom answers, but nods his head, receiving, somewhat passively, her various bits of information. The son is looking no longer at the distance before him. Though his head is erect, extended on his neck, his eyes are fixed on the places before his body, watching the nature of the ground on which he puts his feet. He is thinking about miracles. He is thinking, grimly, about the bride's face. He knows in a short time he will be called upon to do incredible things. This had been a troublesome thought to him in the past, but it is no longer troublesome. He'd known there would be wine in the dipper when he had withdrawn it from the empty cask. His willing it there had been a totally objective act. He'd felt he had no real stake in the miracle, but now he sees that the emphasis, as a possible problem, is elsewhere. He knows, at this moment, that he can do anything he wants. He looks up at the forest on the other side of the meadow. I can rearrange those trees if I want to, he thinks to himself, and then he wills the trees to rearrange themselves. He lines trees up in the middle of the path; he causes the path to grow over with grasses. He strikes all the day-birds dead in the forest. Then in a moment he returns all these things to their previous, favored condition. He knows that what he has done — his literal ability to do it — will never be a problem or issue. But then he thinks of the question asked of the steward at the end of the party. He reviews each permutation of the question, each in the voice of the various askers. He reflects on the face of the bride hearing that question. He fixes the conglomerate postures and expressions of each of the guests at various intervals related to the repeated asking of the question. In each frame he can see the constant lowered profile of good feeling surrounding the asking. He realizes that each guest was either aware or in some way affected by the question. He faces the fact that he is the central catalyst for all these things. He is the cause of the too quickly diminished pleasure of the party. He lets his mind settle on the face of the bride,

45

hearing the two men talking behind her. He measures carefully the slight lowering of her enjoyment. He knows that the day could have been perfect. He knows he is the cause of the slight flawing. He knows the issue has something specific to do with manners, that the breaking of certain forms causes human hurt. He sees the bride's face in the frame; he can see the change in her eyes when the words that are spoken by one of the men enters her ears. The ears take on an almost imperceptible tint of red. "Why did they save the best wine till last?" the man is asking. He knows he must set out to solve and understand, what that question implies.

WALKING

The lands are recovered from the seas and take on their own particulars: diverse shapes coming forth out of eruptions, tensile strength from distillation. And metals build chemically under the ground, and above the ground a texture so various it becomes impossible to put one's foot down in the same place. Hard, smooth rock, corrugations so rudely defined they will gash the toe that comes in contact with them. Soft grasses on the other hand, and inland sands giving way gently to mark the outline and weight of each step. And the seas that remain when the land rises up seem to remain somewhat similar. Usually flat, though possibly rippled by waves, that even though they come irregularly, are measurable. Inland, the lakes arriving ultimately in the same way, and accepting the same offerings. The waters, even as the lands, be they murky and dank or porous and clear, often as crystal.

He is said to have come down from the mountains, and in neither version is there evidence that he stopped, perhaps to consider, on the sand. But rather, he is said to have walked right out to them, shrouded, most probably, by a darkening of the light. The sea, in one version, was not calm: *the boat was in the midst of the sea, buffeted by the waves, for the wind was against them.* In the other telling, simply *a strong wind was blowing.* At any rate they did not exactly see him coming, but rather *beheld* him walking to them upon the water. And what he said, and most probably because he saw they were afraid, was: "It is I, do not be afraid." Whether or not that answered the fear, had it been put in the form of a question, is another thing. Anyway, they desired to take him into the boat; and immediately the boat was at the land toward which they were going.

1

He might be walking along the road near a small Arab village; as it is he is the principal of a high school in Bisbee, Arizona. At a distance of twenty miles the lights of the town cannot be seen, as the foothills that turn finally into desert stand between them and the highway. He is walking along the highway, Route 80 from Tombstone. Ostensibly, there can be no reason for his being there.

At a distance of half a mile, at a right angle to the highway, the rabbits are already moving. Yuro, the leader that has no conception that he leads, leads the bounding jacks — like flocked birds or the dark shadow of a school of fish — along an inevitable line to the crossing. From this distance, given a possible light, it would look like a monstrous blanket rapidly waved a foot over the surface of the desert. The wind they create catches in the cups of his ears with a rush.

The lights and the car itself reach them in the following manner: Mr. Wilcox is caught for an instant at the corner of the beam; he is recognized; he is dressed in leather riding breeches and wears a monocle at the end of the loop of a silver chain in his left eye; in his right hand he holds a riding crop, and with each stiff-legged step he smartly smacks its barbed tips against his leg; he walks briskly; as his feet strike the ground puffs of sand, like dust or water, curl from under the soles of his boots; Yuro, the leader that has no conception that he leads, is killed by the left fender striking him; at least seventy of the five hundred rabbits strike, or are struck by, the car; the car finally stops; some jacks lay dead by the road side, while others flop wildly in front of the car; the car is covered with bits of blood-soaked fur and flesh; the warm smell of it is in the air.

2 — the first dream

In the first dream he stands at the edge of a lake. The water is so calm that, although it is dark, he can see the flopping of fish jumping reflected in the lights of the city a half mile across the water. Mr. Wilcox helps him into the boat, then stands on the shore watching as he rows quietly out and into the center of the lake. When he reaches the center he drops anchor, being careful to keep the prow pointed toward the place on the shore where Mr. Wilcox was standing but now no longer stands. He surveys the shore line to the right and to the left. The lake is surrounded by pine trees that at most places come down almost

48

to the water. There is no beach to speak of, but rather a few grassy knolls that extend no more than a few feet into the lake's edge. Jesus stands on a knoll a quarter mile to the left. At first he says to himself, though he knows it is truly Jesus, "It is a ghost; it is Mr. Wilcox in disguise." And then immediately he feels a great longing to talk to the Savior, to have him come to him by walking the distance between the boat and the land. And even as he feels this longing, the Savior dismisses the fish who had come in the meanwhile out of the water to rest their panting bodies on the knoll around his booted feet. And looking out to the boat, the Lamb steps onto the skin of the lake at the water's edge. Immediately the boat is at the land toward which it was going.

3

After the weekend and the cleaning of the car, which requires a great quantity of kitchen cleanser and hours of scrubbing (and still there are flecks of flesh on the bumpers and bits of fur caught between the metal teeth of the grill), Peter returns to school. And he returns with a changed attitude toward his principal, Mr. Wilcox. The change is one of emphasis, for much of what he feels, though previously only hinted at in the background in ways too subtle for grasping, was nevertheless always there and no doubt operative in its influence.

Mr. Wilcox, because of his large size and spindly structure and his habit of rocking back and forth on his toes like a perpetually idling automobile that might suddenly pull away and in any case is only stopped temporarily in constant anticipation of motion, has always been an object of derision for the students. They make faces behind his back, imitate his walk and rocking motion, say Mmmm Mmmm in the way he constantly does with just the proper intonation, and in general consider him a fool. Peter, being (and wanting to be) no more than an average boy, has always joined in these antics. But now he feels differently, and when, on that very morning, he encounters Mr. Wilcox in the hallway rocking, Peter feels that when the principal finally does take a step forward something will begin, something that he, Peter, is to be in some way a part of. In addition, and in the ensuing days, he begins to see that Mr. Wilcox is in a very important sense a kind of final court of recourse. He seems to stand as an ultimate Judge in matters that affect Peter's life.

In the first stage Peter and Yuro, Peter's friend, in some way manage to get into situations that cause their being ordered to report to the principal's office on many occasions. As often as not they are innocent of the misdemeanor slated against them. Once they are accused of lighting a firecracker in the halls outside the office of an aged female asthmatic. In this case they are guilty. At another time they are thought to be the perpetrators of a dubious joke that involved putting a dead and bloody rabbit in the grand piano of the music department. In this second case Peter is innocent, but he is not quite sure of Yuro. In any case, what is most puzzling to Peter is the attitude of Mr. Wilcox on such occasions. The principal sits behind his desk, rocking back and forth, and demands, regardless of the boys' protests of innocence, repentance for what they have (or haven't) done. He seems, thinks Peter, not to care whether or not they are guilty, but only seems to want them to be sorry that such a thing has been done at all, regardless of who might have done it.

Peter is more successful at answering the principal's desires than is Yuro. This is because he *does* feel ashamed to be in the principal's office, and guilt or innocence seem to have very little to do with his feeling that way. Yuro, on the other hand, argues and protests (always in direct relation to his guilt or innocence), sometimes like a mad rabbit thrusting himself frantically against the perpetual engine of Mr. Wilcox's intractability.

In the second stage Yuro, the righteous, is killed by an automobile while running, for no apparent reason, in the desert.

4 — the second dream
The second dream is no more than the following schema of the first:

Legend: 1 — Peter
         2 — Jesus
         3 — Mr. Wilcox

5 — the third dream

In the third dream the boat is heavy with fish to the gunnels and sits very low in the water. Some are ribbon fish, reptilian, exotic and ugly, with heads of barracuda and the malleable bodies of eels. The others are barracuda, halibut, fluke, garibaldi, blowfish, perch, sand shark, mackerel, tom cod, sea bass, pickerel, squid, dogfish, bream, hake, smelt, albacore, scrod, porgy, lungfish, hammerhead, whiting, sardine, moonfish, opah, manatee, lamprey, herring, and flounder, and drum. Peter sits very carefully in the center of the boat a half mile off the beach on the coast of Ensenada, in the midst of the sound of the panting fish slowly expiring around him. The boat bobs freely though gently in the rippling water of the large bay. As it turns in its jagged circle, Peter sees the edges of beach marking the extent of the bay and then the seemingly endless vastness of the sea that opens from it. To his left, as the boat turns, is a small sandy island, little more than the spine of a bar humping out of the water a quarter mile from the shore at a right angle to the boat. On the scant beach of the island stands Jesus.

The hair of the Savior is gray, close-cropped, yet long enough for the breeze to lift it slowly waving in the air. The Lamb wears leather riding breeches. But his feet are bare, and around them furry forms, rolling and hopping so small and close to the

51

ground at this distance that Peter can only guess by their movements that they are rabbits. Feeling the boat drift once again to
head its prow toward the main beach and away from the island,
Peter is seized with a feeling of mixed fear and longing. He
stands up precariously in the boat full of fish and gazes over his
left shoulder. Jesus dismisses the rabbits, who hop and roll over
the gradual sloping sands of the island and disappear from
sight, and then, looking at the wash of the surf a few yards in
front of him, the Savior steps deliberately onto the water. As he
walks toward Peter, and though he is graceful and delicate, he
appears like a man walking on mattresses. With each step his
foot sinks a few inches into the skin of the bay, giving just a hint
of buoyancy which at another time and in other circumstances
might appear ridiculous and awkward. As it is Peter is afraid.
Perceptibly the wind rises, catching the flaps of the pants of the
Master, who is now halfway between the island and the boat,
causing the new waves to rise up around him. As he continues
to come the wind he might be said to have created catches in the
cups of Peter's ears with a rush, whispering: "It is I, do not be
afraid." Immediately, and with a lurch that sends Peter flopping
among the panting fish, the boat begins to move. By the time
Peter can regain his feet, the boat sits on the sand of the land
toward which it was going.

6

The desert is not known for its friendliness to outsiders, and
those not yet particularly versed in its ways through evolution
seldom find a place for themselves in it. For this reason, if for
no other, the jack rabbit is an interesting animal. The desert is
foreign to the rest of the land. It is stolid and intractable, and
yet, like the sea, it constantly shifts its contours like something
awake and alive. Jacks are warm-blooded, eat vegetables, and
cannot find sustenance or shelter in the desert. Yet jack rabbits
are very plentiful in the barren Pacific southwest of Arizona and
New Mexico. They live in those parts near farmland oases where
they can find their fill of fresh crops. But often there are so
many that, having exhausted crops in one place, on occasion the
necessity arises to cross bits of the desert itself in order to get to
fresh food. They understand the desert; they know it has no sol-

ace to offer them, and when they pass through it they do it together and as quickly as possible.

As for Yuro the righteous, he knows that because he is swiftest, he leads. The five hundred sit at the edge of the plowed field, the beating of their nervous hearts humming like some gigantic perpetual engine. It is Yuro's place to give the signal to leap out, to explode all of them out into the intractable wastes before them. What will start him is possibly a change in the wind bringing an odor of greenness to him from a great distance. The job will be to go in a straight line from this to a point somewhere in the distance as quickly as possible. All that remains is to start.

7

In the third stage Peter tells Mr. Wilcox that he has seen him walking along the highway in the desert whacking a riding crop against his leg. This occurs on a day when Peter has been called down to the principal's office for coming to school barefoot. Mr. Wilcox is rocking and humming behind his desk. Peter, his feet uncomfortable upon the rough wood of the floor, sits in complete acceptance in a stiff chair on the other side of the desk. "Let us be honest with one another, my boy." And Peter tells him of how he had taken the car keys before his mother had left for her job on the night shift at the telephone company; how he had waited until she had time to walk the short distance to her job, and had then taken the car away from the scene of his fatherless home to rush, for reasons that he did not fully understand, into the barren desert away from the town. He tells him about his speed and the wet sound of the hot asphalt under the tires, of Mr. Wilcox himself, caught for a moment in the light beams, and of the bloody and flopping rabbits, the perpetual rush of their motion and breathless panting in the boat around him. It is then that Mr. Wilcox comes to him.

8 — the fourth dream

Peter, the leader who has no conception that he leads, works at securing the lines and the nets of the boat. It is already dark and though they are over halfway to Capernaum, Jesus has not come to them. Now the sea is rising, because a strong wind is blowing. The wind catches in the cups of Peter's ears, and in the way that such small sensings will often do, it reminds him of his youth in the not too distant past; he remembers how he would lie and dissemble, how he was somewhat wild and sad, and how he would often leave the strangely foreign town of his birth to throw himself alone and mindless against the stolid intractability of the surrounding sea. He is afraid for what he was. But then he remembers the coming of the Master so suddenly down to the shore saying "Follow me" and how he had dropped his nets and followed him. He thinks of the hand, not the rough scaled hand of his father, but the soft pulsing palm of the Lamb, resting upon his head; of that same hand, and the delicate arch of the Master's foot that he himself had held, and not like a fish, so soft and pliable against his own rough fisherman's palm. And then Peter has no fear or longing and reclines with the others, who are still somewhat afraid, around the gunnels of the boat. The sound of their quiet breathing goes out over the water.

And in the fourth watch of the night he came to them, walking upon the sea. And the rest, seeing him walking upon the sea, were greatly alarmed and exclaimed, "It is a ghost!" And they cried out in fear. Then Jesus immediately spoke to them, saying, "Take courage; it is I; do not be afraid." And then Peter answered him and said, "Lord, if it is thou, bid me come to thee over the water." And he said, "Come." Then Peter got out of the boat and walked on the water to come to Jesus. And though he saw the strong wind, he was not afraid; and he did not begin to sink. And soon he had come to Jesus. And Jesus at once stretched forth his hand, taking the cheek of Peter in it, and drew the fisherman to him. And Peter felt of the body of Jesus, and felt the toes of the Lamb mingling with his own like the nibbling of small jeweled fish in the water under the surface. And then the Master kissed him. And Peter felt the lips of Jesus and knew in his mouth the taste of the sweet, saving, tongue of the Lamb.

And then, after a while, the two, still locked in their embrace, walked in the soft calm water back and into the rough wood of the boat. And immediately the boat was at the land toward which they were going.

54

LOAVES & FISHES: A PARABLE

There are fat men everywhere in the world, rolling around in it, making their way by air, or heavy as ballast in steam ships lumbering between continents. He knows this as well as the rest of us. They walk often painfully, with thick legs wrapped in elastic bandages, their feet slowly collapsing in their shoes; they lard the earth. Or else they don't walk at all, are carried from place to place by others who, being less fat, can handle the job. And make no mistake about it (he knows this as well as we) everybody loves them. It is not that they sing and dance, as in the false old saw, nor take up, of their own accord, the job of entertaining us. It is rather the force of the words themselves, "fat man," and their exaggeration of our own place in the world.

"How man reasons?" he thinks to himself, in the parable, as he stands in the deep meadow, here among 5,000 people reclining on the green grass in groups of 50. There is first, simply, the thin man who has read the insurance company reports. They tell him the fact of his own thinness adds years to his life. Then there is another, similar kind, loving the fact of existence in someone else (thus, not in himself) of what he considers obscene. Last is the hungry man, whose love is quite literal; the word becomes flesh to him: possibilities of steak and fatty chops, of rich shank, barbecued ribs, and the thought of plump hunks of sweet meat imbedded in fat along the spine.

"Gather up," he says, "all of the fat men and bring them to me." At once, at a distance of 100 yards and from over a rise in the meadow, come 2 thin men carrying signs. To the left of them, 2 more, also carrying signs. They come to him slowly, almost failing under the weight of cardboard that tends to wave heavily over their heads. The first 2 approach and stand before him. The signs read "fat" and "man." They sit down with a plop in front of him. No one is talking. Then come the other 2, who also sit down with a plop. The signs read "everybody" and "loves"; he applauds them. The sound of his soft hands goes out over the heads of the silent people sitting in groups of 50 on the green grass. "Gather up," he says, "all of the fat men and bring them to me." At once, over the same rise, come 40 thin men swaying under the weight of their burden of 20 fat men. Even before they have set the fat men down in front of him with a plop, the people — who are beginning to unfurl banners and open bags of confetti — have understood him, and in the rising commotion only a few hear his statement: "And the word was made flesh (make no mistake about it) and dwelt among us."

Picture 5,000 people, most of them thin, rising up from the green grass of a meadow in groups of 50, their waving of banners and throwing of confetti, their singing, and their applause. Picture a crowd of this size with 20 fat men rising above it somewhere in the middle on 40 thin shoulders. Picture these fat men, draped with streamers and bright-colored robes. (There are also flowers in their hair.) Everybody loves them. There is also some envy and drooling and licking of lips. There is also some gnashing of teeth.

1 — the lesson of history: his own voice
"Take 5,000 Indians and 5,000 Historians. The Indians are at a powwow, or are preparing for battle. Possibly they are dancing an Indian Dance around a fire. They might be listening, attentively, to their Indian Chief. Whatever they are doing, they are doing it Historically; the Oldest Indians see to that. Of the 5,000 Historians the Majority are Students of History. A few of the rest are History Teachers. The Rest are watching what is going on. The Students and Teachers might be found in a powwow. They will be discussing, of course, History. Should they begin, however, to dance around a fire somewhere, they will be considered Mad in their tracks. There is of course an Historical Precedent. Should 5,000 Indians and 5,000 Historians go into Battle many on both sides will be Killed. There will be some, however, watching from a Distant Hill. There is a Lesson to be Learned from this. It has something to do with History."

2 — loaves & fishes
In the parable he goes among the people and produces loaves and fishes. There are two accounts of the miracle. In the first

version he suddenly stands up among them as if to speak. He purses his lips in the attitude of a fish and raises a heavy arm above them. The rolls of flesh below his mouth appear as loaves of bread. He gradually lowers his arm in a slow sweeping motion, indicating the green grass beside where the 5,000 people recline in groups of 50. Lying on the grass beside each person is a loaf and a fish. In the second version he says to them, "How many loaves have you? Go and see," and when they have found out they say, "5, and 2 fishes." And then he says to them, "Give me 1 loaf and 1 fish, and distribute the rest evenly among the people." Then they distribute the loaves and fishes, and the people eat them. And when they are finished — and in both versions — there are 12 baskets of fragments left.

Now when he has finished his 1 loaf and 1 fish and he looks on the grass beside him, he sees that there is still 1 loaf and 1 fish left, so he eats them. But just as he is finishing the others come up to him with 12 baskets of fragments. He tells them to disperse them again among the people lying in groups of 50 upon the green grass, but when he looks to his side he sees there is still 1 loaf and 1 fish left. After he has eaten them and has discovered that there is still 1 loaf and 1 fish left, he stands up among them and says: "Gather up all the fragments, and bring them to me." And after they have gathered up all the fragments, which amount to 1 loaf and 1 fish, they bring them to where he was sitting. But he has gone from them in order to disperse the fragments.

3 — the meeting

He prepares to go to the meeting; he wakes up in the morning. It starts that early. He thinks to himself: "I have five hours to prepare," but immediately it is noon. Coming in from the island he thinks of the few words he can say, and how only a very few of these will be heard. Everyone will have make-up on and moulages of wounds in rubber covering exposed parts of their bodies. They will have make-up like chalk on their faces and false fangs and false hair. But some of it will be real, real people among the others. And the trouble is that he won't be able to distinguish between them, and will have to remember to keep within the small circle of lights that even now is prepared, waiting for him, humming. Crossing the bridge he remembers an old vision: a man leaping from a moving conveyance, diving over the bridge beams, down 100 feet into cold water. He has thoughts of a valley surrounded by soft rolling mountains like breasts 2,000 miles to the west. Now he can see the building itself. It is rubbery and atomic among the other hard forms. It even pulses. He knows they are sitting in their make-up, waiting. He enters the door.

4 — on the grass in summer

He sees them lying on the grass in summer. He can see no anger in them. Their feet are gorgeous; even in such awkward shoes he can imagine them running, as if they were barefoot, gracefully up to him. Only on such days in summer can he think clearly, and keep track of the lessons of history. George Washington might have stood on similar ground, wanting to call up similar children to him. The square teeth he had in his head were part of the common wood; he, also, was one of the people. And Kit Carson, and Davy Crockett, and Tom Paine. And even the sad-eyed Indian leaders, such as Sitting Bull.

He sees them lying on the grass in summer. He has become famous. He has invented loaves and fishes. He cannot touch them. He, too, has become abstract. And so he leaves them, attempting, gracelessly, a trot. He, like George Washington, wants to be called, in kindness, by his first name.

5 — the message

In front of him are two microphones. He seems to be listening to one of them. No, he is reading a small paper. The way he holds it, it could be a Chinese Mystery Box. It is very fragile. He holds it delicately on the tips of his fingers; it could be written on tissue. By his expression, it could be a telegram: *tragedy. stop. L has died. stop. come at once.* By his expression, he can't quite make out the writing.

He is surrounded by microprones; at his back they are held by newsmen. He is surrounded by bright lights.

He holds in his fingers a small paper. No one is talking. But everything around him is electric, and everything hums. His chin rests on the knot of his scarf. His head leans to the side. Everyone is waiting.

He holds a small paper that could be written on tissue in his fingers. He can't quite make out the writing.

A SUSPICION OF CRUELTY

When we mention the purity of his vision we are not speaking of morals. It is true that his eyes would burn, sometimes with a fire like that from blue stones in a dark cave, under his strong forehead, set in the perfect symmetry of his face; also, that his hair would seem to burn and curl, as if charged with electricity, and would keep moving, around on his head, indoors, after he had returned from a walk on a particularly windy day. People were often dumb-struck in his presence. He could see into things, into the heads of the people, into the pure properties of their wills, could speak, even across a room, so that the listeners would hear his words like cool breath in the middle of their heads, as if he were close to them, whispering in their ears, and nobody else in the vicinity could hear what he was saying. But not morals: at least not of a personal quality. He loved people of course, and each in that exquisite personal way, close to their own need for him, spiritually. Nothing was missing in his life.

But the days are becoming eventful; he is walking, locked in the terrible magnitude of his own existence, to where they are waiting for him, desiring him, the word he has told them that is his body; for him to give it to them; he believes in its crystalline purity; this is the source of his problem.

Now he comes to where they are waiting: a place of a jumble of rocks in the foothills. He has told them, lately, to write things down: his simple pronouncements; he can't understand the ways they can confuse them, when they repeat them, tell them to others.

> No one lights a lamp and puts it in a cellar or even under the measure, but upon the lamp-stand, that they who enter in may see the light. The lamp of thy body is thy eye. If thy eye be sound, thy whole body will be full of light. But if it be evil, thy body also will be full of darkness. Take care, therefore, that the light that is in thee is not of darkness. Therefore, if thy whole body is full of light, having no part in darkness, it will be illumined, as when a bright lamp illumines thee.

60

A Lesson from a Lamp; and they take this and talk about darkness in cellars, the way one keeps a lamp alive, though buried under a bushel, the coming of thieves in the night, the way the body becomes a dark temple when the eyes are closed. They look into his eyes, and then they speak of them (thinking that he can't hear them), calling them lamps, perpetual flames, used also for warming dark cellars, filling up bodies with light. He has no conception of irony.

And so he provides them with books to write in, a size they can carry in their robes, but with firm boards, like slate (though light) ; and they can spread them firmly over their knees, like tablets, and write as he speaks to them. Little red books, that amaze them, that are perfect, covered with fabric; they have never seen such books; or bookmarks, woven so tightly, as smooth as silk. And they write in them, and quote from them. He has seen them, raising their fingers into the air, talking to small groups of people, reading to them from the book.

Always, when he approaches them, they are holding their books in their laps, and there are always new ones among them, who hold awkwardly their new books.

She is sitting on a rock at the back of the crowd. She sees him approaching. She is holding her new book in her hand by her side; the spine rests in her palm, which is sweating; but the spine of the book is cool, is dry, and her sweat doesn't soil the spine; the spine dries up the sweat.

And as he approaches, she can see the rip in his garment. He has mended it with twine, but there is still a space there, and she can see the skin of his thigh, the white skin above the tan-line, where the sun has darkened his legs from so much flopping of his robes as he walks. His body is erect and moves only forward as he walks. He places each foot on the ground firmly, but always toe first, as if he were stepping into water. His head doesn't bob as he walks, but remains at perfect plumb with his feet. And she had heard of his eyes and his hair, but to *see* them: his eyes like literal lamps or burning blue stones, his hair churn-

ing on his head, a particular closed system of rhythms, each
curl in its own way, moved by its own private wind.

The land of a rich man brought forth abundant crops...
See how the lilies grow...
A certain man had a fig tree planted in his vineyard...
Strive to enter by the narrow gate...
When thou art invited to a wedding feast...
A certain man had two sons...
Salt is good...

They write these things down. Or he asks them to read back
to him the pronouncements that meant the most to them in the
last session. He sits on a large rock in their midst, his legs folded
under him. And often they read back things that are misunder-
stood or vaguely formed. He feels enough to allow them their
own voice, at least in inflection, sees that they must make at least
some little translation in order to make the pronouncements
their own, to possess him, to take him into their half-formed
bodies. "What do you feel about that?" he says: "Don't ask
what it *means*; in what circumstances would you speak it? To
whom?"

He has, of course, realized her presence at the back of the
group, though she bends over her book and does not partici-
pate. Even as he was approaching them earlier, he had spoken
into her head, in a way that she could not hear as literal statement: "Welcome." He is sure that she feels welcome. Bends
over her book and waits till he speaks, and then she writes constantly, all the time he is speaking. And he allows himself to
imagine, his words forming into statements of crystal in his
mouth, into crystal balls or glass birds, made out of fragments
of glass in his mouth. The fragments are the letters of his words;
they form together; they rise out of his mouth like a glass bird,
like a puzzle formed into glass, and there are no lines of suture.
The body of the bird is completely transparent; it does not take
space in the air. It moves from his mouth, goes and hovers over
62

her book; it breaks into pieces; it falls, and forms into letters and words on her page.

And he keeps on talking and working for clarity, and twice, for a moment, he feels they can glimpse what he means when he says, "I am the Word." And then he finishes for the day. They mill around, reluctant to leave him. He calls her to him, saying, "Daughter, show me what you have written today." She comes up, shyly, to him; she turns the book around and shows him a drawing of his face.

He works at the kiln, not firing pots, but cleaning the place up. Shards of broken pots, dead wood and ashes from the last firing, even her easel, standing, beside the kiln, the hinges gummed up with paint. He'd been a real potter when he was a child, working beside his father, who made pots for a living. He saw it as hard work, not fun or whimsey. This damn mess, another object she'd brought to the place, used twice or a dozen times, and then discarded, like the dog she'd got for herself really, not for the kids, and now he had to feed it and brush it out. And plants, and artful oil lamps, drawers stuffed with junk, furniture, constantly rearranged to give her places to pose for some strange cult or other, and that given up too, after a while, but the dragged-in, painted, and disheveled furniture remained.

A man comes to himself in the middle of his life, finds he is in a joke, is either the hero or the butt of it: seven inches in snow or out in the literal cold. One time she'd been devoured by some ridiculous cult that met in the desert. The cult was itself a joke to the town. "You goofy Bildock," people would say to each other, using the cult's name as cliché for people who walked around in a daze. She'd come home, abstracted, her clothes disheveled, the make-up smeared on her face. And so he had followed her one night, to a place in the desert, had watched her take off her clothes, climb up on an altar, have intercourse with a man in colorful robes. And then he'd jumped out, when he couldn't stand it, saying: "What are you doing to my wife?!" and someone had said: "Don't worry; he's giving her

63

resolution." And the line had come straight from the middle of his head; he could hardly contain it; a joke within a joke: "Thank God. I thought he was fucking her." And that had been the end of it. He refused to question her. When a man finds himself within a joke, he is either silent or he plays his part. He cared for his three children; he worked when he could; she spent the money; he spoke openly, only when she was out of the house.

More shards, more globs of paint, even a scrap of paper, under a foot of the easel: some silly bit of a poem. And going up to the house later, he sees her approaching. In this transmutation of the joke she is wearing simple clothing. She carries a red book pressed against her breast. She reaches the house before him, and when he enters she is already sitting, huddled on a straight-backed chair in the corner, looking intently, into the book.

Somewhere, someone is making the books. It was not necessary, but he allowed himself the indulgence. They were in his room, wrapped as a shipment; perfectly made, they are uniform. The book has a red cover; the corners of the book are rounded. The spine fits tightly. But when the book is opened it rests flat in the lap, becomes almost rigid. You can start at one side and write clear across its face; there is no discernible break in the middle of the doublepage. And when the book is carried, when it is placed under the robes, it stays where it is placed; it needs no pocket. It is always dry. Any instrument can be used to write in it. You could write in it with your finger, but nobody realizes this; they would think the idea a joke. He has no conception of irony. There is nothing on the cover of the book; each book is exactly the same. Everybody knows their own book when they touch it.

They had come to the time of their first private meeting. She had asked for it, saying: "There are things in the writing I'd

like to talk to you about, in private; could we...possibly...
meet somewhere?" And he had invited her to the room he now
used. It had a private entrance, faced onto a garden, was very
cozy.

And now he sat with a bundle of papers in his lap. These are
the writings, "And I'm not sure yet; I don't want to enter them
into the book till I'm sure. When you speak to the group, it's
another thing; the words just seem to form on the page, but
when I get home, and think about you and what you've said, I
just can't get it together and write it down."

He sees she is dressed very simply. She wears no make-up he
can discern. She is probably ten years his senior. When she
speaks she speaks very flatly, as if the words meant nothing at
all. But the fact of her coming to see him is a serious matter,
and therefore her words must have meaning, that he must
search for, somehow, around the words. He looks at her face as
she speaks; it seems on the verge of expression; when she finishes, he looks down at the paper.

She has written of things in the parables, things barely mentioned, things that are beside the point of the lesson. But not
of lamps under bushels. The Just Steward, for example, who
betrays the rich man, comes back to the household and visits
the rich man's wife. In one place she has written: "The Master
speaks. His words fall like a blood bird from his mouth. The
bird hovers over my book; it breaks into bits, like drops of red
crystal semen; it falls on the page; I kiss the book." "What were
you reading just then," she asks him; he shows her the paper
he'd just read; she smiles. "I get the point," he says, "you write
very clearly, but your words seem to be translations of something other than what they say. Remember, I am the Word, and
I shall enter you, but you must become open to me; from your
own life, from your own place, the years you have lived. Remember, ten years ago: the clothes that were worn? The songs
that the people sang then? What was called beautiful in women?
And how that has changed? I remember those days, and how I
was fixed in them; they were beautiful; they still live for me;
I long for those times. But the words that are spoken for the
book are transcendent, pure in their vision, transparent as crystal. They are not fixed into any time."

Then he tells her to take back her papers, think on what he has told her, and next week they can meet again.

A man is the butt of a practical joke; or a man is a teller of a joke; or a man is a simple listener. But jokes are ironic, and if anyone laughs at a joke, there is someone present who finds himself at least on the edge of being in the joke, as either teller or butt, or the hero or the man passed on the road in the prologue, before the joke begins in earnest.

She is back in her room in the house where she does things that he thinks are possibly jokes. He cleans at the house; he picks up her plain garments, which she has discarded over the beaked head of an animal icon of some forgotten cult. He scrapes at the dishes; he tries to avoid touching any object that is hers. Ten years of madness.

And then he hears her come out of her room. He turns, and before he sees her he catches her shadow, strange and familiar, against the far wall. And then she appears. She is dressed exactly as she used to dress ten years ago. Instantly, he loves her.

She sits in the chair in his room. She is dressed, precisely, in the style of the fifties. There are seams down the backs of her stockings, which are slightly coarse. She wears a powder-blue skirt; it fits tightly around the hips, but is long below the calf when she stands up. Her only license is the split up the side of the skirt to the thigh. It is open an inch, and is held together by crisscrossing bits of gold like metal imitations of leather thongs. He can see the tops of her stockings, as she crosses her legs; the tip of a garter is visible. Her hair is long, but turns under, in a page-boy, at the shoulders. A white barrette, in the shape of a white bow, is pinned to her hair on the side of her head, holding it back and in place. Her lips are painted dark red, and the same red glossy color covers her fingernails, up to the half-moons of the cuticles, which are starkly white. High spiked

66

heels, with open toes; the two toes protruding on each foot are also painted, but are seen more vaguely because of the thick toes of the stockings. When she leans forward to hand him her book, he can see a good bit of cleavage; her breasts are elevated, protruding from her loose-necked peasant blouse, lightly powdered; on the left breast, in the groove near the sternum, is a large, dark, artificial mole.

What she has written for him is odd, but very direct this time, emptied of all but the literal, but back in another time. Phrases like: "When the merchant again returned to his home he was agitated and put his hands on his wife's boobies; the prodigal son set out on his journey to find some thigh; the wine steward had hot pants," gave evidence both of directness and distance. "You are getting there," he says, "you are getting there." But she seems bewildered, and squirms in her tight skirt in the chair. "Don't you understand?" she says. "Isn't this what you wanted?" "Yes," he says, "but remember, even the prostitute in the streets of the city, if she shall come to me, if she will open herself to direct speech, I will enter her: she will possess the Word."

The next time she came dressed as a prostitute.

He follows his wife down the path at a distance. He does so against his better judgment. But lately she has become more frantic, tearing among her objects. Last week she emerged from her room dressed as a prostitute. She keeps writing meticulously in the red book, sometimes for hours, and almost always she tears out the pages and burns them. And so he follows her. He has heard her whisper the name of Jesus; he has heard of this man called Jesus, creating somewhat of a stir in the town, the leader of another cult, but one that seems outwardly harmless, without ritual or obscene display.

When he reaches the place where his wife has stopped, at a jumble of boulders in the desert, he pauses, stands back a little, out of sight, and watches what happens.

The man in the center of the ragged circle is obviously Jesus.

He sits on a high rock, his legs tucked up under his body. He speaks to the people around him.

A young man came into the city to increase his fortune. He was a male prostitute. He began by working on the streets. But after a while he was discovered, and was taken into the finest brothel in the city. Here he would dress in women's clothes sometimes; he would take men into his body dressed as a young shepherd; he would enter women while pretending to be a bull or stag.

There were various rooms in the brothel; each was totally distinct. Sometimes clients would take him into many of the rooms in the course of their visit; he would always behave in distinct ways as these clients moved him around, but the ways he behaved were not necessarily directly connected to the rooms he was in.

Then it came to pass that a merchant, a man of warped sensibilities, entered the brothel one night. He also had come to the city to increase his fortune, had sold his wares for a profit, and was now ready to fulfill his conception of pleasure.

The merchant carried a satchel with him, and he refused to leave it with the clerk when he entered the brothel. He demanded to be shown to the room of the young male prostitute, for he had heard of the young man. But not one of the distinctly different rooms of the brothel, but the private, personal, real room that the young man lived in. And because all in the brothel feared him, they took him to the young man's room.

Now the young man's room was indeed his private place. He kept in it only those objects he deemed necessary for living his life. When the merchant entered he saw that the young man's room was bare; the young man stood in the center of the room; he was dressed modestly, in robes of plain fabric. The merchant's warped sensibilities were shocked.

"And is it *this* that I have come all this way to discover?" the merchant screamed at the young man. "Where are the devices, the clothing, the engines?" And the young man produced from under his garment a small pillow, of silk, with tassels at its corners, that was embroidered with a map of the various rooms of the brothel, their location, the colors and objects and decor of each.

"That is nothing!" the merchant screamed; and he took from his satchel a small, complex, intricate machine. He placed the machine on the bare floor of the room.

Now the machine was a small replica of the brothel itself. It was like a dollhouse with the top removed; all the rooms were visible from above. And in the rooms were tiny replicas of human beings, all poised in various positions, just prior to the fulfillment of lust. The young man could see that in each room there was one figure that was himself. The merchant toyed with the switch, like that for an overhead lamp, that was set into the side of the structure. And then the merchant looked up, in order to smirk at the boy, in triumph.

But where the boy had stood there was only empty space. The boy, himself, was lying on the floor, his robes open, like a blanket under him. His head rested on the silk, tasseled pillow; his feet were together, almost touching the machine. And in the middle of his body, where his chest had been, there was a place like a hole in a jigsaw puzzle, and the shape of the hole was exactly the shape of the merchant's machine.

And the merchant quickly attempted to replace the machine in the hole in the chest of the male prostitute. But the shape of the hole had changed, was indeed closing, and the machine would not fit, and the boy was dead.

The merchant left money for the boy's burial.

And when Jesus had finished there was no discernible sound in the group, and each member was gazing intently at the book in his lap. Then there was the sound of tentative laughter, and the man quickly realized that it was himself who was laughing, and he couldn't stop. So he quickly withdrew himself. And all the way home he kept chuckling, breaking even into laughter again, but wondering, wondering about the nature of the joke.

The book is in no sense magic. And the words that are entered in the book do not perform tasks. The people read from the book. They shake their fingers in small groups; they produce the book at the least provocation; what is magic is the way it appears in their hands. The book itself remains solvent. It is static, does nothing, is no bond of community, and yet it is beautiful, cool to the touch, is protected.

If all the books were gathered together in a place and put to the torch; and if the ashes were gathered and placed in the satchel of a merchant, who when he had traveled a good distance began to succumb to the weight; and if he then took the satchel and weighted it with stones and cast it over a high cliff into the sea, all talk would seem to end with that act. But the satchel would rot where the stones rolled by the sea rubbed against it, and the ashes would enter the water, rise up, and float on the surface of the sea. And then the ashes would be fired by the sun, as if in a huge kiln, into crystal, would gather together, forming the shape of a bird, of a size that would take four miles on the surface to rise; and rising, the bird would displace all the air, and when the people below were gasping for breath the bird would break into pieces and fall to the earth, and everybody there would be rearranged.

This is a sense of the possible power of the book. The people quote from it, shaking their fingers. When they are going someplace, it rides on the skin of their sides, under their robes; it needs no pockets. It could, just as well, be elsewhere.

This is the sixth time she comes to see him in his room. She sits in the same chair; she is wearing a mixture of clothing; her hair is a variety of styles. There had been times she had come wearing subtle make-up, but this time her make-up is erratically applied. Once, when he had mentioned the body as a sweet temple or herb pot, she had bathed her body in oil and scents, had come to him in the fragrance of a flower garden. Another time he mentioned the positive qualities of dirt, our need to get back to the soil to which we all shall return someday. That time she had not bathed for the whole week, had even worked in the garden with her husband, had come back to his room smelly with sweat, and dirt was visible in the creases of her skin. But the last time there had been nothing that he had said to aid her, though she had searched for the words that would guide her in a way to please him. She had gone half mad during the week be-

70

tween. Searching the book and her closets and chests, she tried on various postures and clothing. She had bathed herself in oils and animal dung. Since he had mentioned "The Little Children" in his talk, she had even tried to find some way of capturing the scent of her own children, by rubbing against them, sleeping in their small beds, wrapping their dirty clothing around her body, under her robes. But she knew she had failed, and now she was sitting, she thought as a failure, before him.

This time her writings were erratic with passion. She had written directly into the book this time, and he held it, open, in his lap. "What are breasts for but to suck," she had printed inside the outline of his left cheek in the first picture she had drawn of his face. She had drawn other pictures. In each one his face appeared, but in various sizes and expressions. Sometimes his face was surrounded by drawings of female organs, ones that were specific, obviously, her own. There were old jokes in the drawings: breasts, marked *sweet* and *sour* beside their nipples; there were male organs also: penises, called *prod* or *infernal machine*. In one place she had traced a picture and writing from a small pocket-sized book; she had titled the page *Tijuana Bible*. The picture was of a man with a woman lowering herself onto his swollen member. In a circle, indicating the man's speech, she had carefully printed the following statement: "I am your engineer on the locomotive of love; climb upon my jumbo-sized joy stick, and I'll teach you all about fuckin' in the big city." She had altered the couples' faces: she is the woman; the Master is the speaker of the words.

It was then that he realized the sheer force of the distance that he had taken her. She was so close now to the opening. She had moved erratically through the various trials. She had bracketed with metaphor; her passage through the specific period of the fifties had been the most dramatic exclusion of her real self from the process; he had had to be very careful in dragging her back from that movement. But the last time he had given her nothing to hold to; he had forced her back to her own objects, had allowed her the vain search. And now she had presented herself to him, literally *dressed* in the dark seeds of her frustration.

She sits before him; she is like the new bride who comes to

the wedding with the totality of her past self worn on her virginal body: something old (in this case the white plastic barrette in the shape of a white bow, the summation of her own real past, which ends at this point); something new (the gathering, in her erratic make-up, of the various forms of her struggle, now breaking into fresh configuration); something borrowed (pictures and words from old jokes); something blue (catalytic, the aura of her whole body, in the light of his eyes). The new bride comes to the altar. Possibly she is shy; her hands fumble in her clothing, and then they rise up like fluttering birds that come to rest, cupping her breasts, for a moment, then turning over, the palms up. It is as if she had removed her breasts, as if they were transparent, invisible chalices of glass that she is offering to her new husband, so that he might fill them, with blue wine. And this is the literal gesture of her own hands now; her forearms rest along her legs (which are slightly parted at the knees), her palms turned upward, resting on her knees. And then she speaks, saying, "Master, there are things in the writing . . . I . . . ." And then he knows that she is ready; he distills each posture of time, takes it out of time, fuses it together, and returns it to the vessel of the small room in which they are sitting. And then she begins to speak openly to him.

She tells him about the various operations she has had. She gives him her age and the ages of her children. Her husband, she says, is fat, has a colostomy, washes his body incessantly. She tells him she is the one who has caused her husband to do these things. Her two children are growing into homosexuals; she admits she likes that, and enjoys watching the strange sexual way they ride their bicycles. Sometimes, she says, she sleeps with the children's dog. She tells him of the many cults she has belonged to, the things that priests have done to her body; but there is one cult that she still believes in, and she is afraid that the Great God of this religion will destroy her aging body for loving him, Jesus. And yet, she says, she *does* love him, that she can attach no strings to the quality of her love, that she will give her body to him, even though it is a poor body, and will surely be crushed into dust by the Great God if she gives it. And then she weeps.

72

But Jesus distracted her from weeping; he spoke crisply to her: "Look at this trick!" he said. And when she looked up she saw he was holding her red book in his hand. And even as she watched the book began to spin on his open palm. And then the book rose up into the air, spinning ever swiftly. And when it was spinning violently, suspended in the middle of the room, it began to change its color, from red slowly to blue, as if caused by the increasing heat of its friction in the room's air. And then he said, "Look at me!" and as she looked into the blue flames of his eyes, he extended his right hand, placing it in the air, directly in the middle of the beams of their gaze. And then he snapped his fingers together. There was an incredible cracking sound in the room, and then dead silence. And as he removed his hand, a slow shower of flower petals began to fall over their bodies, and then they both looked up. And though they were no longer in the room, were instead seated under a massive flowering tree, growing in a wide meadow of grasses, she was not shocked or surprised. And then again he looked into her eyes, and said, "Here I come." And as her hands turned over on her legs, her knees slowly closed together, and she accepted him.

He had been following her for a long time and a great distance. She has been moving like a madwoman, keeping close to buildings, her hair disheveled and stringy. Now he rounds a corner, and she is gone. But he keeps on walking, until he comes to the edge of the town. And after he has gone even further, he comes to a large hill and climbs it; he finds himself looking down into a wide deep meadow. The meadow is full of long grasses and wild flowers; there is a brook running through the meadow. His wife sits under the tree, leaning against the trunk; Jesus sits on the grass facing her; flower petals are slowly drifting down from the tree, over them.

Suddenly the man feels he no longer lives in a joke; he is not on some periphery of what he is watching, and therefore no butt or observer or listener. The meadow is a closed system, and he is no part of it. Even the fact that he is watching it does not include him. In an old joke two drunk men emerge from a bar after the sun has risen. They begin to argue as to whether what they see is the sun or the moon. A third man passes; he carries a lunch pail; he is obviously on his way to work, and sober. The two drunk men stop him. Perhaps he can settle their argument. "Tell us," they say, "is that the sun or the moon up there?" 'Don't ask *me*," the man says, "I don't live around here." And he feels he is like that man, really, who has no place in the joke, is catalytic only, like the coin that the two drunk men are searching for, but not where they have dropped it, but under a lamp post, because there is more light there. The coin and the man: the first is the necessary condition of the joke, the second enters only to provide the conclusion, then leaves just as quickly. What do the two drunk men do with the answer? Is the coin still in the gutter? Did the joke, in any sense, exist?

It does not make sense to even speak of him as being on the edge then. His wife, the man they call Jesus, are sitting under a massive tree. They both stand up and begin to walk, very slowly, across the meadow. They do not touch each other, but they walk very close to each other. When they reach the narrow brook, Jesus reaches out and takes hold of the arm of the wife; then he aids her across the brook; then he releases his hold on her arm. That brief action could not be described as intimate or personal, nor had it been simply thoughtful or pragmatic. It had nothing to do with crossing the brook. It had been exquisite and pure, so completely familiar as to transcend any possibility of interpretation. It is as if he had taken her arm as one takes up a book; one carries the book with the spine resting along his palm. And the book has been written by the man who carries it, who does not remember what is written in it. Because it is the perfect book, it is a literal extension of the man who carries it.

But then he thinks: "He is after all helping my wife across a narrow brook, and the way he takes her arm seems to fit no possible context for such an act. He is not helping her, guiding her; there is no term of endearment present; he is not acting the gentleman; and what would it mean to say, even, that he has taken her arm?"

74

And then, as the master releases the arm of the man's wife, and they proceed slowly, down to the inverted groined cup of the meadow, what the man feels grows into possible statement. It is that he feels, in the way Jesus has held the arm of his wife, that there is, after all, something nameable. He suspects that there is something cruel in that man; and then, that in that there is also the slim possibility of a new joke. He is wrong on both levels.

CORRIDORS

He remembers he came to the town in a boat. In the late summer; in the afternoon; a bright sky, and he could see buildings a mile off, of various shapes, the sun hitting them. They seemed to come clear to the water's edge, had been built the moment the people had landed, as permanent shelter. But they were big, and as they came closer, he saw in their shadow the place where the wharf had been built: in front of the buildings, a place where the boat could dock. If it were to dock, but in this case it wasn't. It would pull up straight to the pier, turn at the last moment, sidle along the old boards, and he would have to jump (possibly two feet) over gunnels and onto the landed wood.

They'd talked about this. His mother had said: "The boat will pull up straight to the pier; it will turn at the last moment, sidle along the old rubber tires, and you will have to jump, about two feet, over the gunnels and onto the wood of the dock." "Won't my legs push the boat back when I push off?" "No, the boat is too big to be moved like that; just jump when I tell you; you'll be all right."

That he had not come to the town by land was somehow important, marvelous still, to him after seven years. He'd go in a rowboat, sometimes, out to a safe distance, and then he would head in quick and try to reclaim the feeling. But rowing, he'd have to look over his shoulder, and he couldn't go as smooth or fast enough, and it was never the same. Besides, he had begun to see the town from over a mile out to sea, and it wasn't safe to row out that far.

He was six years old; the center of his experience of the town was now at a point a half mile of water away from it. And the wharf had begun to form; people were sitting on chests on the dock; someone was moving and selling something; to the right was a fisherman, casting his net; the dots of old tires, a few gulls sitting on pilings, swirls of oil in the water. "What if the boat's wash pushes it back out for a minute, just when I'm jumping?" "I'll watch for that; I'll tell you when." "Will somebody catch me if I start to fall back?" "Yes; don't worry about it now."

76

He had always thought it as strange that he knew the town best from a different element, a literal place over water, from the prow of a boat. It would have been one thing to come to the town over land, following road signs, riding or walking. Even if he had been alone, and not with his mother, he would have come to it gradually; it would not have seemed such a solid place. The way one walks from the straight, planned areas of the suburbs, into the twisted streets of the city: the change is so gradual; the city only slowly gathers around you; but even if it were not for that, the ground is the same ground. But to come to it over water: the city has almost a line that defines it: the wharf, the outermost pilings, the immediate sea.

But earlier on, in the ship's hold, sitting beside his mother (her arm round his shoulder); nobody else, but the seamen coming below, at various times, to check them; some being gentle and kind. But you could never tell when the light got cut off, that shaft down the passage, a corridor, through which came some man, maybe the one with the leer, maybe the one who brought water; head in his mother's armpit: "How far are we now?" "Only a little while, a little while yet."

1

The City is a real city: the people live in the streets or in houses, lining the streets, facing away from them. The streets are running with garbage, with animal offal, the side streets running with human excrement. No gutters, but corridors carved by the rain pushing through cabbages and shit, to someplace indistinguishable. The City could be said to rage and seethe, or it is foolish to speak of the City as raging and seething, for the City is not human: is not people, garbage, or houses.

He comes to himself on a street in the heart of the City. He is unsettled by his own presence there, as if he has come to meet an important person. He tells his apostles to leave him, and then he takes up a post where the street turns into a small square at the center of the City, and the people begin to approach him.

First are the women with young children: limbs that he touches with both hands, the mothers kissing the hem of his robes. And then come the blind and the lepers, and one man who cannot speak, and Jesus touches his tongue with his own spittle, and the man leaves him uttering nouns. A man comes up to him and says: "This is what I did in life," and he begins to dance and tumble, sometimes walking on his hands, sinking into the muck of the street up to his wrists. And spins on his toes, dragging his left leg, like a thin stick, but grotesquely swollen and arthritic at the knee, which suppurates like a rotten cabbage. And Jesus touches the knee and cures it, and the man walks away firmly, no longer playing the acrobat, in a stately posture. And the rest come, each in their place and order.

And then after a few hours of curing, he begins to reckon some similar faces returning. "Why do you come back again?" And the man who had danced says: "The pain has moved from my leg to my stomach, and I have to walk bent over." And Jesus touches the man's stomach and cures him. And others come back: a man who'd been blind now feigning dumbness, motions to Jesus to touch his tongue with his spittle, and shudders when the Master does so. And the lepers return with headaches, and Jesus can hear people, going away, talking of the relative pleasures of the various cures: "When he touched my tongue with spit, I could barely stand it!" And then he begins in earnest. He is curing a man who has returned to him six times, and each time with a different ailment, a different location and intensity of pain. He had come first as a leper, dragging a pus-filled leg, and Jesus had touched his leg and cured him. Then he had returned in an hour, a deep pain in the middle of his chest. The third time his eyes bulged, seemed ready to jump from his face, and Jesus had touched his eyes with spittle, and they had receded. And now the man is standing before him. He has lost considerable weight, he complains of a pain in the head, and Jesus walks carefully around him, touching the various parts of his body; he measures his limbs with a bit of string; he measures the circumference of his head, and then he stands facing the man; and then he strikes him full in the face. The man rocks back on his heels; the pain sinks through his chest to his stomach; he becomes incontinent; the pain escapes from his body in feces and urine and runs in its own rivulet, into a corridor in the street.

2

It is at least possible that the ship will not get close to the
wharf. Two feet is very close; it's a big ship. Even four feet is
close, and I doubt that I could jump that far. But if my mother
stood behind me and gave me a good push, the arc of my jump
might take me that distance, at least get one foot on the edge of
the dock, and someone would surely grab me. If my father were
here, he would build a board, like a gangplank, about twenty
feet long (the width of the ship). I'd help him attach it to one
side of the ship with hinges. We'd use heavy wooden pegs to
secure it. While we sailed it would rest across the gunnels. And
then, after the ship had turned, in order to sidle up along the
wharf, we'd flop the board over, so that it would hit the dock,
and I'd jump up and run down it, onto the safety of the wharf.

3

Why do you stand in the street? Why do you allow them six
visits? What is the nature of the message? Is the City a human
being? Why do you cure them singly? Why must you touch
their bodies? Why do you distinguish women? How could you
send us away? How did we get to this City?

A house is offered, and Jesus takes them into it and through
it to a bath at the rear; the bath provides the livelihood of the
owner of the house, but he gives it over to Jesus, who has cured
his son, for nothing. The bath is narrow, and along one side of
the room is a long bench, and Jesus tells his apostles to remove
their shoes and be seated on the bench.

They sit in a row in their robes; some cross their legs; others
sit with both feet together on the floor. Their sandals have provided little protection and their feet are covered with offal and
various dirts. Jesus can tell, by reading their feet, where they
have been.

He takes up the first foot in the line and begins to wash it.
He uses a cloth and a small brush. Someone has provided a
basin of warm soapy water, and he moistens his cloth occasionally. He washes with exactitude, but quickly, but as he proceeds
he slows down, lingering over each foot for a long time, washing each foot in a way particular to its anatomy, shape, and
structure. Peter is last in the line, and when he reaches Peter's

feet, Jesus takes time to consider. He does not touch Peter's feet right away, but directs Peter to lift each foot and turn it, so that he might better look at it. And then he takes up the left foot of Peter. And Peter feels the Master's hand like some incredible fine shoe, one he knows he could walk on forever; the Master's palm forms to his arch; his index finger divides his large toe from the second, but feels like a strap of velvet formed to the space between them. Then Jesus proceeds to wash Peter's feet. He produces a small brush from his garments and carefully scrubs at the cuticle of each nail. Where there are spots of cracked dirt between the toes, the Master bends and moistens the place with his tongue, so the dirt softens and runs to the floor. Peter squirms on the bench, constantly moving his head in order to always see the Master's hands, his face, his pink tongue, washing the dirt away. "This is Jesus," he repeats to himself, "he is putting his lips on my feet; my feet are the farthest point of my body; it is like he is kissing my face." And then Jesus rubs at the callouses with a piece of mortar, goes over the heel, the ball of each foot, the tips of the toes; he pares down the nails with a small, sharp knife; he takes each toe into his mouth and swirls his tongue around it. And as Jesus moves his tongue down and between each toe, Peter can feel, in his own mouth, as if his teeth were the toes, and the Master's tongue is moving across his gums, slowly, identifying each tooth separately, curing the roots, moving on. And Peter feels that his teeth are breaking into words, that from now on each tooth will have a separate and distinct movement when he speaks, so that his words will have a new, incredibly complex generative power of inflection, that he will never say the same thing twice again. And even then, as the Master holds his small toe in his mouth, their eyes locked together, Peter can feel the first real sound of his life. It rises up from his stomach, passes his larynx, takes particular shape in his mouth, and enters the room as a moan. And each object in the room changes; the room becomes — literally — a new place. And when Peter looks back at the face of the Savior, he sees he is smiling, his bared teeth sunk into his toe; a trickle of blood is running down his exquisite chin.

4

He remembers carrying his father on his back, walking down steps, and placing his father in automobiles. Sometimes he'd only be taking his father out of the house for air. His father never liked to sit in a chair in the yard, but wanted to sit right down on the grass itself; and so, at these times, he would have to walk to the edge of the house, allowing his father to hold to the side of the building, to slide from his back and rest in an awkward way in the grass.

His father was not heavy, was easy to hold, and yet he was always afraid to carry him. If he had thought of his father as a permanent cripple, if they had at least spoken of him in that way, he might have mastered the act, been able to carry his father in comfort. His father would turn in his chair, away from the table, and he would face away from his father, bend at the knees, and await his father's hand on his shoulder. He knew his father was grimacing behind him, working to lift himself, a little off the chair, using his crutches, both in one hand, to aid him. Then he would feel his father's hand on his shoulder, would reach round behind him, grasping his father's bony buttocks. His father would release the crutches; someone would take them, and then he would wrap the bone of his fore-arm under the chin and across the neck of his son, and the son would rise up slowly, letting his hands slip from his father's hips down to the backs of his knees, and would stand up, the light weight of his father upon his back. The left knee of his father was swollen the size of a cabbage; the lower leg on that side hung like a piece of thin tin along the thigh of the son. He'd hold the crook of the right leg in his wrist; he'd hold the left knee like a cabbage resting in the palm of his hand. His father usually rested too low on the son's back, but he dared not hitch him up, for this was an awkward position for his father, and any movement would cause him pain. He always feared he would fall, and would literally break his father in two, or would fall on top of his father, crushing the almost empty cup of his hips; would crush his delicate ribs with his own spine, falling upon him. He imagined how it would be, to leap up from the bed of his father's body, turn round and see his father's face as he cursed him through the pain of his last breaths. Once, while carrying his father, he misjudged the width of a doorway, had banged his father's left knee into the side of it. His father had cried out in pain, cursed him and began beating him feebly on the head and shoulders. He'd been so surprised he'd dropped his father's right leg, and felt his father begin to slip. But his father had grabbed tight to his throat with his left arm, had kept cursing and striking him, saying "Go on! Jesus Christ! Go on!" and he had gone on, weaving from side to side, down the stone steps, had turned, backing his father into the seat of the car.

Sometimes his father would become incontinent at the table. He always watched his father try to eat, and could tell at such times, as his father's face would begin to sweat, his teeth grind together; his eyes would widen, in effort to control his sphincter, losing the battle: the sound of the gas, the shuddering of his father's body, the gurgling sound of the shit leaving. When only the family was present his father could withstand it, would say things like "Here it comes again," would manage a feeble smile. But when guests were present, his father could not hold back his humiliation; his head would drop to his chest, and he and his mother would carry the chair with his father rocking in it, away from the table and into the nearest bathroom; and he would leave them, giving his father over, to the secret care of his mother.

He remembers they carried his father into a room. They were good friends of his father, old school chums, and his father had lost some bet or other, to them, and they had picked him up, laughing and joking, as when people are picked up before they are thrown into a lake, and had carried him into a room in the house, had closed the door after them. When they carried his father in he was large and robust.

He did not see his father again for a long time. But he slept in the room next to his father's room, and he could hear things, if only vaguely, through the walls. Often his mother would go into his father's room, and he would hear the sounds of the bed squeaking, limbs striking against the wall. Neither had his father's chums come out of the room, but one of them would occasionally call to him from the door and send him to get things from the store: a rope, a hammer, pieces of wire, two metal rings, and occasionally baby food.

And then he himself fell sick, with a fungus infection that completely covered both of his feet. He lay in the adjoining room, both feet out of the covers extended over the edge of the bed. They had put towels on the floor under his feet, and the towels became gradually matted with pus and scabs. The feet themselves were like two rotting indistinguishable vegetables, that seemed to shudder and breathe with a life of their own. He could not sleep, fell into delirium, was in constant pain. And then in the middle of the night he saw his father standing in the doorway. He was not sure how he knew it was his father, for the figure his father had been was now bent and emaciated. He was naked, and the son could see there were various devices attached to his father's body. Both of his father's wrists had been pierced with hooks, and his left knee had a thin electrical wire running from it, with a plug affixed to the end. His father still wore his belt, but the belt was now so large that it could have encompassed two fathers, and his father had to hold onto the belt in order to keep it from falling into a large circle at his feet. "My son," the father said; and the son leapt out of his bed, staggering on his ruined feet, but before he could reach his father someone jerked his father back into the other room, and before the son fell to the floor he glimpsed, through the closing door, an incredible device like an upright rack, affixed to the wall of the room.

5

Now I remember a time before the beginning of the journey. I am still sitting with my head tucked in my mother's armpit; we are still on the boat; but I remember a time in the Midwest, in which I have broken my clavicle. My father sits in a white summer suit on the porch of our house. They are tearing a side of the building away, to put in a picture window, facing a crab apple tree. My room faces the top branches of the tree. But I am sitting on the porch with my father. A violent electric storm is beginning. And there are old photographs and

83

movies, in which I am standing or moving, in the park across from the house, under the corridor the trees make, beside a plaque of the names of the war dead; or I am running, in a soldier suit, with a model airplane in my hands, across the lawn of the park, to where my father sits waiting in the grass. And then I am walking the four long blocks from school with my broken clavicle. I begin to mimic the new limp of my father. And as I am moving along, Lady comes out of the hedge two hundred yards away; she turns, and faces me. The sidewalk is narrow, lined all the way along with hedges on both sides. Even at this distance she is larger than I am, and there is no escape. She is running, and what I remember are the last few feet of her run: the way I am guarding my arm with my books, the fixed shape of the dog in the air, the slow arc of her trajectory.

I don't know who's running this ship; I imagine my father somewhere in the workings of it. It moves now in a straight line, as if through a trough in the water. The distance is shrinking, the dock beginning to form itself. I see the figures of animals and humans. Let it be two feet away. Let it hold for a moment.

6

He is twelve years old. He is dressed only in underwear. His mother is carrying him on her back. They pass over the threshold and out of the room. Someone quickly closes the door behind them. His feet dangle like rotting vegetables. His mother speaks to the crowd who has gathered: "Look, I have given birth to a son!"

7

If they had come to the gates of the City, before the shipwreck, still in the faint guiding light of the late afternoon. If Jesus had not washed their feet. If the streets of the City did not run with vegetables and excrement. If it were not called Babylon. If the pople lived in the City and did their work and when they got sick were cured and remained so. If no man offered his house. If the streets of the City were not narrow corridors, but straight and as wide as twelve men in a row. If Peter had not
84

demanded to be carried; if he had understood the parable. If
they had not lost their way and come at last to the dock.

8

And so he had become his father's child, though born out of
strange circumstance. Finally, they had brought his father out
of the room, had placed him in a wheelchair at the kitchen
table. The wheelchair would not fit through any opening in
the house; it was not collapsible, and yet surely they had not
built the wheelchair inside the house or built the house around
it, but at times he saw the wheelchair outside, in the grass, his
father resting comfortably in it. At night or in the afternoon,
when they had placed his father in bed, he would push the
wheelchair around the house, looking for exits. He would bang
it against doorways, try lifting it through windows, but it would
not break or collapse.

And then he began to fall sick often. And during such times
he would beg his mother to bring the wheelchair into his room,
saying vaguely, "It gives me a certain comfort." And he would
lie awake with his ear crusted with scabs over suppurating
wounds or his feet throbbing with draining advanced fungus,
and would look at the wheelchair, would sometimes reach out
and touch it from his bed, would get up and rub himself
against its firm metal. And sometimes, late in the night, he
would think that his father was under the bed that he slept in;
he would fancy his father was thinking of him, would imagine
his father reaching out, his arm moving across the floor: his
hand closes around a spoke; he pulls the wheelchair closer to
the son's bed.

But the son remembers a day when they had taken his father
and placed him in the grass beside a large walnut tree in the
front yard. They were talking, but his father was not listening,
was looking up into the higher branches of the tree. And then
his father took a pencil from his pocket and a scrap of paper,
and placing the paper on his knee, he wrote, upside down and
backward, so that the son could read it directly as it was written: *let's get some nuts.* And then he began to carefully help his
father, who constantly cursed him, "Christ, be careful!" until

his father was standing, holding onto the thick trunk of the tree. And after he had helped to lift his father into the lower limbs, he stood back, beside his mother, who had come out of the house and was wringing her hands. And his father slowly made his way into the higher branches, cursing and moaning, shaking the tree as he moved. And the nuts began to fall to the ground, and among them the son could see occasional pieces of fingernail, small bits of skin, drops of blood, spirochetes of hair. And when his father had reached the highest limb, which was the richest of all, he gave it a violent shake, and a hundred nuts were thrust down their shafts to the ground. And then his father rested, suspended in the branches, like a spider, his body came to stasis. And looking through the leaves and branches the son found his father's face, and his father smiled, and the stigmata passed into the son's palms.

9

A quarter mile out, and now he can see how the people take shape into occupations. There are people simply sitting and waiting. The fisherman hauls in his net. A man sells nuts, baking on a steaming cart. He holds close to his mother's body; they are standing in the prow of the ship. And down from a road that ends at the dock itself, he sees a group of men, who proceed carefully; they are intent on watching where they put their feet. One of them is carried, by two men, who hold him as if he were seated in a moving chair. His feet are brilliantly white, and they hang and sway at the end of his legs like large delicate honeydew melons.

"Two feet," he thinks, and turns his face to his mother, for the ship is now coming to the dock, is beginning to turn, in order to sidle along the old boards and tires, and he will have to jump, possibly two feet, over the gunnels and onto the wood of the dock. "Will they catch me if . . . ?" and his mother assures him they'll catch him.

The ship turns slowly; the approach is accurate; the thirteen men seem bewildered at finding themselves on the wharf. He stands, with his mother's help, on the gunnels. The stern of the ship overshoots the arc of the intended turn; it crashes into the corner of the dock; a large hole is ripped in its underbelly; water rushes into the voids. One of the thirteen men extends his hands and beckons to the boy. The boy squats on the gunnels; his mother grasps him by the buttocks; he thrusts himself into the air. The stern of the ship overshoots the arc of the intended turn; the prow overcorrects: it crashes into the side of the dock; the dock begins to buckle; water rushes into the forward voids. The boy slips on the gunnels, falls back on his mother; he crushes her body beneath him; water rushes over the deck; the chalice of her hips is splayed and broken; she curses at the son as she drifts away. The ship comes in at a perfect angle, sidles along the old tires lining the wharf; the boy squats on the gunnels; his mother holds onto his buttocks; everyone rushes to the side of the ship to wave good-by; the weight throws the ship out of balance; it begins to list; the boy leaps into the air; the superstructure crashes around him, splitting the dock into pieces. The boy rushes across the breaking dock; he reaches the mouth of the street; the twelve men gather around him; the thirteenth man is missing. The ship rocks away from the dock; the returning wave of its wash speeds it back out to sea. His mother stands in the stern of the ship; she holds her veil around her face at the neck; she waves back at him with her handkerchief.

10

And now have I come to the Fabulous City. I float over it. A mile from shore, in a boat, a twelve-foot skiff, that does not rest at anchor; the sea is so calm and flat. I float in a circle, my prow raised by the weight of myself in the stern; turn, past the sea and the other city at a distance, empty. No clouds. No real sun. The blue dome; the inside of a blue egg. The flat sea. The slow turn of the boat.

And under me the City rests in the care of the water, through which I can see, as if through a clarity of crystal. Fish, and long vines and sea grass at the edge of the circle, in which my boat floats, turns on the viscous surface; slightly convex, my boat like a dark pupil in the center of a still blue eye.

Float. Turn. And below me the superstructure, the barnacled

gunnels of the huge ship, as large as a city, changed in a new element: starfish, sand dollars in rows around it on the sandy bottom. The whole thing seems to list; the board across the gunnels rises at its loose end, in the currents, and falls back with no sound.

Sinews of loose beams and guy wires, tendons and ligaments of cable and rope, I know now, which are the parts of my father's body, that have not been rotted or wasted, but wrested, from the possible hands of morticians, into usefulness, into specific potency.

And the caulking a mixture of clarified body fat and diseased tissues. And the braided cables are his tendons. The ropes are oiled and protected from rot with synovia and mucus. And the chinks in the injured voids have been cured with his tissues. And over the deck of the wheelhouse, his stretched skin. And under the glass dome of the compass, his brain.

My boat turns, and turns quicker. At the edge of my circle, the beginning of a whirlpool, turning slightly; my boat is beginning to descend. And the sky becomes electric. And storm covers the city; a black cloud, I can see as my boat turns, obliterates the city. Torsion of the boat turning, now quickly. Nothing but water around me.

And now I can see as my boat lowers; I look over its side, down in the clear water; my boat slows in its motion; it hovers at a level with the highest mast, turning ever so slightly, above the ship. A wheelchair waits on the deck below me, a blanket, folded, over its metal arm. And then the door of the wheelhouse opens; my mother comes out of it; she carefully descends the ladder, carrying a chalice in her free hand.

And now the City rises to meet me. The boat stops turning, descends, past superstructure and wires, then comes to rest on the deck.

And my mother comes up to greet me then; she hands me my crutches, on which I raise myself, and helps me to sit down in my chair, and aids me in folding the blanket across my knees. And then she hands me my chalice and kisses my cheek.

And then my mother looks up at the wheelhouse for a moment; she nods, and then she turns back to me. "Hold tight,"
88

she says. And with a slight gentle jerk, the ship rocks free of the bottom and rises; and soon our prow breaks the surface; the ship emerges; the water spills down its washed sides and rushes over our ankles. And for a while the ship rests and bobs on the flat surface. And while the sun is drying our clothing and hair, we talk of our lives together. We are tender. We laugh often.

And then my mother looks up at the wheelhouse again, and then she turns and looks back at me. "It's time to go," she says, and reaches and locks the wheels of my chair in place, and then she holds tight to the chair's back.

And I am sitting in my chair in the broad prow of my ship; my mother stands behind me, one hand touching my shoulder. We look out from the prow, into the flat empty surface of the sea. And then the ship plots out its course. His brain rotates under the glass dome of the compass. And the City moves away from shipwreck.

PASSION AND DEATH

THE LAST SUPPER

Now had they come together, in a narrow room, on the second floor of the house of a friend. Jesus had said to them: "Gather the proper foods." And Judas counted out coins, and two others went out and bought things. The table was properly set. There was a cup and an empty chair, close to the open door, so that Elijah could enter. And then they began the Supper.

And when the cup was raised, and one of them stood by the door, they heard a sound in the stairwell, a clump of feet, and then Hound entered the room and leapt up on the empty chair and rested his forepaws on the table.

No one was shocked at the entrance of Hound, for he was the Master's dog, and often sat by his Master's side while they ate, and the Master would openly feed him scraps, from his own plate and from the plates of others. But to witness Hound as he sat in the chair of Elijah, his paws on the edge of the table, to consider the possible blasphemy, and then to see him lower his muzzle to Elijah's cup and grip the cup by the rim with his teeth and tilt his head back, wine draining over his teeth, into his whiskers and down his throat, these things shocked them all, and they looked at the place where Jesus sat; they were dumb-struck and questioning.

But Jesus smiled, and looked at the face of Hound, who sneezed out the wine that had wet his nostrils, shook his head violently, then gazed in the eyes of the Master. "Now has he gotten to you too, old fellow," the Master whispered, and none of the rest understood these words. But Jesus could see in the eyes of the dog that the devil had entered his body.

Now there *is* a devil. And the way the devil gets to us is not mysterious. This is taught to people, especially to children, who

91

are told about it, so that they will wind up being good. The devil makes things right. Whether the things are good or bad, it does not matter. Every time a man opens a door, for instance, the devil is there; he helps him do it gracefully. If the man is entering a room where there are women, ones whom he wants to impress, the devil guides his hand to the knob, makes it turn, like the bolt is swimming in oil, guides his foot over the threshold: the grand entrance. Imagine if the devil did not do that: the man trips as he enters; his coat gets hooked on the doorknob; the arm of his suit rips off; he is wearing a short-sleeved shirt. This goes for everything we do. We are fragile. The devil knows that. He uses it against us.

Hound sits at the end of the long table, and Jesus looks in his eyes. There are tears running from his own eyes, and the dog that is deep within the form of the dog that the devil possesses sees this, and gives a low howl of pain from the pit of his stomach, and shudders in Elijah's chair. "Fear not, old fellow," the Master says, "for when I sit by my father, your head will rest on my knee." And for a brief moment, Hound takes up residence in his own eyes again, and his mouth curls up in a smile, and then he is swallowed by the devil's presence.

The first time Jesus had met the devil, he had come to him dressed as a merchant. Indeed, Jesus had heard that this merchant might come to his rooms to see him. There had been talk about a rich merchant coming to town, looking for various goods that he might buy, export, and reap a profit. It was said that he was particularly interested in fine wooden bowls and spoons, and Jesus, in deference to the craft of his father, often spent time working with various woods, in carving bowls and spoons. But Jesus thought of himself as a kind of Sunday artist; he cared not to sell the things that he made, and did not care whether or not the merchant came to see him.

But he had come; they had talked prices and quality. "My, but you make fine bowls," the merchant had said, and the very bowl he had handled had splintered into pieces in his hands. "You have broken my bowl," Jesus had said. "Nonsense," the merchant said, "there was a flaw in it." And Jesus had ceased to speak, and merely kept looking intently into the merchant's eyes. And after a while of this the merchant began to shuffle his feet, to act generally uncomfortable, and then he spoke.

"Look," he said, "I am the devil. Don't misunderstand; it was a little joke, and a kind of object lesson. The flaw wasn't in the bowl at all; it was in you. But it wasn't really a flaw. It's just that I took my powers out of the bowl. It's those powers that helped you make it right." Jesus said that he didn't believe that. "Look," the devil had said, "would I have allowed you such a strong name if I didn't like you? 'Jesus Christ,' it has character, music in it; the right name for an artist!" For the second time Jesus did not believe what the devil told him about the bowl. "Come on," said the devil, "look, you're right, breaking your bowl was a gauche way to bring it up, but I can give you things, and I just wanted to give you a hint of my powers."

And then the devil took Jesus out of his room, and flew him to high places: the top of a mountain, the highest building in the town. He promised Jesus many things; like riches, women, good wine, freedom from sickness, the pleasures of flowers, mysterious unduplicatable wood tools, a wife, a measure of wisdom, satisfaction, visions, a long life. And all he asked in return was that Jesus fall down and adore him. But Jesus remained constant, and finally the devil gave up, and then he went away.

And now Jesus was facing the devil again, in the seat of Elijah, locked in the body of a dog. But not any dog, but Hound, the close companion of Jesus.

The apostles squirmed in their seats; they leaned to one side; no one spoke; the seder had stopped. Hound sat still in his chair; his tail twitched through the slats, his forepaws rested on the table; his great head faced the Master; his eyes burned with fire. And Jesus saw he could no longer melt the gaze of the devil; he could not speak to the dog in him.

And then the apostles heard the voice of the Master. It entered their heads almost as though they had not heard it, but had remembered it, entering into them, then being recalled, a moment after the words were literally spoken: "One of you will betray me." And immediately the apostles began to murmur among themselves, and Simon Peter spoke in the ear of Judas, he whom Jesus loved, who was reclining on the Master's bosom: "Who is it of whom he speaks?"

And then Jesus did the one great thing he could do in this situation. He smiled, directing their gaze back to the chair of Elijah; he placed the total burden of guilt on the devil. And though the devil tried hard to direct their gaze back to the Master, though he rooted his body around in the chair, thrusting his muzzle in the Master's direction, he was unsuccessful, and all the apostles' eyes were upon him. And then he shook his head and growled, and with his right paw he struck at the cup of Elijah, knocking it over. And though they had thought it was empty, a trickle of wine spilled from the cup; it bubbled like sherry, and traced an intricate line like the path of a dying snake's movements over the table. It was not soaked up by the cloth; the line traveled toward the seat where the Master sat, turned at the last moment, and spilled in a trickle, into the lap of Judas. And he whom Jesus loved jumped up from his seat, one hand holding his wine-soaked crotch; the other covered his mouth. And he ran from the room in horror. And the apostles rose, for they were angry, and wanted to strike Hound and beat him. But Jesus stayed them, saying "Shut fast the door, and let us continue the meal." And then he smiled at the devil, knowing he could not escape.

And the devil tried to extricate himself from the body of the dog. And Hound shuddered, and with the help of the Master he held the devil in him, and the devil was frustrated and angry and afraid of humiliation.

And then they continued the seder; they began to sing. And as their voices rose in the room, the devil began to shudder again, tried burying his muzzle in his fur. But Hound jerked his head free, raised his muzzle in the air and joined them: he sent a tuneful howl into the room, mingling with their voices. His head was extended, but his eyes were the eyes of the devil, and they turned to the Savior; they burned with incredible fire, they spun. And smoke came forth from the dog's nostrils as he sang: smoke from the cooking, frustrated heart of the Devil that burned with his own crucifixion. And then Jesus rose among them.

He gathered some pieces of bread and a large chalice of wine. The devil was spent in the dog's body, and sat, slumped over, in Elijah's chair, his head hanging upon his hairy chest. And then the apostles saw, for the first time in their lives, the possible fires of vengeance able to burn in the City of God. The Savior stood in his body like a piece of ice; the father merged with the son; the tongue of fire of the Holy Ghost hovered, with frightening heat, over his head; the body of the Master was possessed, but not by diluted power. An idea rang in all of their heads: *this is what is meant by purity.*

And then he gathered the pieces of bread and passed them to each of the men, and he poured wine from his chalice into the cup of each. And never once did he take his eyes from the figure of Hound in the chair. And then he spoke.

"All of you take and eat of this bread, for this is my body." He held a piece of bread in his hand, and as he spoke he cast it into the air of the room, but it did not fly fast, but began to float on the air. And the horrified eyes of the devil came to rest on the bread, and he began to cringe and moan in the voice of a dog. His hair stood on end along his back, and the bread moved and rotated toward him. And it was changing. It turned like a piece of bread on a floating spit, and as it moved it began to transform itself into flesh: subcutaneous fat, with a slab of skin on one side, he saw in his fear as it turned; and rich red meat, with dangling veins and arteries; strips of muscle and ligament; fascia and peritoneum, on the other side. And then Hound forced the head of the devil up from his chest. The mouth of the dog wrenched open; his eyes contracted to pin points of fire, receding in fear in his head. And the rich gobbet of flesh floated over his open mouth. And then the Savior opened his robe, and revealed the side of his abdomen. And there, just below the heart, was a ragged opening in his stomach, and as he closed his robe the meat dropped into the open maw of the dog, and the dog's eyes bled as he chewed it.

And then Jesus touched the rim of the chalice and spoke: "All of you take and drink of this wine, for this is the chalice of my blood of the new and eternal covenant." And the wine turned to blood in Elijah's cup, and the dog's teeth gripped it, and then his head sprang back, and the warm blood bubbled in his throat. And Jesus looked into the devil's tormented eyes, saying, "Do this often, and when you do it, remember me." And the dog fell from Elijah's chair, and pulled himself, whining, to the door. "Let him pass out," the Master said, and the door opened, and Hound began to wretch and vomit, gobbets of meat and blood, down the open staircase, and the devil left through his mouth. And then Hound struggled to his feet; he shook himself thoroughly, and then walked stiffly to the Master's side. And the Savior stroked his head, and they finished the singing of the seder.

†

Now everything begins with a meal, and everything ends with a meal. We eat at weddings; we eat at funerals, at parties, at graduations, in small gatherings at the houses of friends. And things begin when we eat. It is like we are taking into our bodies a feast of the possible worlds we will enter, in the terms of the people present, in whatever occasion we eat.

And things end with eating. But not at funerals. For there the food is a start for the living, the first entrance into a new world, formed by the absence of death. We eat the dishes the dead one loved; we eat his share to excuse him. But the thing that ends with eating is the life itself. That's why we say of a man close to death with a fever: he is eaten up with it. It is that the body eats itself; that's why we die alone. We do not cause our deaths; we merely succumb to hunger.

What can we say of Judas? How did he die, and where does the meal come in? He stands at the corner of a building; he leans against it. He thinks to make of that casual gesture a new life, or at least a pause in the one he lives. If he can stand, emptied of hunger and thought: a man out for a walk on a balmy night, who stops to lean for a quiet moment, against the side of a building; if he can submerge his life in a posture, as if in a painting; a man stands leaning against a building, in baggy white pants: he is short and heavy; one hand rests in his pocket, probably running coins through his fingers . . . .

But is is not the lust for coins, nor the feel of them in his pocket; he had no choice in the matter. It is that he finds a small bit of bread in his pocket, among coins, and a small vial of wine. He takes them out, and holds them in the palm of his hand. Both are warm. The bread is fresh from the oven, with a brittle crust; its sweet scent comes to his nostrils. And the small vial of wine! amazing! He holds it before one eye, and everything that he sees is suddenly beautiful. The whole street, seen from the place where he stands, each object, no matter how useful or mundane, comes to his eye in its own integrity. And even the turns that the street makes, the very dirt, from his feet to the other side of the street: amazing, tender, beautiful. He feels like a man, out for a walk on a balmy night, who stops to lean for a quiet moment, against the side of a building.

And then he becomes very hungry. He eats the delicious bread; he tips his head back, draining the vial of wine. And then he feels the building against his shoulder; he enters back into his life. But the touch of the building is exquisite; he can feel the fabric of his rough shirt, between the building and his skin. And then he goes around touching things: he presses his hand in the dirt, he caresses his own body; he presses his lips to the building's side.

And then it all comes marvelously clear to his head, the next few hours, all the things he can touch and see, that will not even end with the exquisite touch of the Savior's lips on his own when they kiss in the garden, both knowing the voluptuous touch of betrayal.

He can see, in his head, the largest tree in the garden, can hear the literal thunder. The sky darkens; the earth breaks open beneath his feet. He drops down slowly from the hanging limb; the alive, voluptuous rope tightens around his neck; it begins eating into his flesh; the tree limb gives a little, and then remains firm.

And he knows even as he thinks these things, as real as they seem to his head, that they are yet to come to him, exquisite, in the very near future. And then he proceeds towards the garden, a constant prayer on his lips: a thank-you note to Jesus, in gratitude, for the gift of the twofold meal of his death.

THE AGONY IN THE GARDEN

On top of the mount is a garden, known for its olive trees. It is a kind of park. A trail winds up from the foot of the mount, which is really a hill, and where it gets steep near the top, there is a small, circular area, a resting place, just forty feet from the garden.

The garden itself is contained in a wall, of mortared stone, about two feet high, but there are open places here and there: the beginning of winding paths, that are lined with stones, that pass through the trees and flowers. There are many stone pools in the garden, and in them are gold fish: garibaldis mostly, over a foot in length; they are heavy and slow, and they often seem merely to hang in the various levels of water. The water that enters the pools comes over delicate falls, like small fish runs, like miniature terraced foothills, and the water splashes down them, babbling into the pools. It is clear that the trees and the sources of water were here well before the garden was fashioned. The paths are delightful; they accommodate to the natural beauty around them, winding between the trees, skirting the babbling falls.

There are many olive trees in the garden, maybe a hundred; the flowers and mosses that grow in the garden are close to the ground and look almost like colored, meandering doormats, dressing the feet of each tree. There is always the sound of water in the garden, but no birds sing.

On the highest hillock in the garden there is a place like the one below, a circular area, a resting place; but here, because of the view, there is a map and a legend carved on a structure that looks like a podium, but only two and a half feet high: the better for children to study it. The map is a contour map of the hills and valleys below; the legend gives names of the various places and turns and dips in the landscape. This is where Jesus kneels, in front of the map (a commanding view). It is early evening, and dusk is beginning.

The garden is not that big, and given the hundred olive trees, it is always heavily shaded. There's a break in the trees, however, and from the podium the valley below is a clear and expansive picture. It is like a picture, in browns and reds, a long way off; nothing seems to be moving; darkness is beginning to gather in ruts in the rolling hills. The valleys are not that simple: the landscape is complex and subtle. The map and the legend explain it. It is getting dark.

This is where Jesus kneels, in a shaft of light that comes through a cut in the trees; it is light from the full moon, now barely visible; flickers of dust like moonbeams bathe his entire body. The rest of the garden is darkening, the trees beginning to look like large and grotesque people; an owl hoots, from a long way off, but the sound seems very near.

An hour before, when the garden was still lively in the filtered light of the sun, and butterflies settled on patches of moss, Jesus had gathered some apostles around him and said: "I'm going up to Gethsemane, up to the Mount of Olives, to pray." "Let us go with you," they'd said, and he had acquiesced, had allowed them to climb with him as far as the resting place, just forty feet from the garden. They had wanted to go up with him, all the way, but he had been firm: "Sit down here, while I go over yonder and pray." And then he had entered the garden, and knelt down at the podium. And just as he had settled himself on his knees, put his palms together, and rested his hands on the map, he heard the sound of feet behind him. It was Peter.

"Master," he says, "we can't stay awake down there; there isn't any breeze; it's been a long walk; we're sweaty and tired."

"Could you not then watch one hour with me?"

"The spirit indeed is willing but the flesh is weak," says Peter. And Jesus sighs and smiles. "Go back once more and try it again," he says, and Peter leaves him.

And now Jesus again knelt down, bathed in the light of the full moon, his hands once again together, over the map and legend, and began praying in earnest. The light from the moon was the only strong light now in the garden. The sweat on his face glistened and mixed with the moonbeams; the trees in the garden now seemed to be moving, brown gnarled outlines, slightly swaying, seemed to extend their branches, over him. A

99

dog raised its voice in a howl in the distance. "Father," he said, "if this cup...," and then Peter was again beside him.

"Master," he says, "it is getting dark now and we are afraid; even Hound is bewaring the darkness; just now he has begun to howl. Could we not.... I mean...," and Jesus says unto him: "Only a moment now, a few minutes. Go back and pray with the others." And Peter, with great reluctance again, leaves him, shuffling back to the others.

Imagine the garden: Gethsemane. By day it is almost bright, cheerful almost. Butterflies sit on the patches of moss; Garibaldis rise to the surface at times in the bubbling pools; the brooks have been formed with respect for the natural surroundings: they rill in the mornings. The paths wind among the trees. But now it is almost dark. The paths and the pools seem parts of a natural forest. The place no longer appears as a garden; this place on the mount no longer controllable and civilized. The place that the map and the legend describe, the valleys below, are beginning to fade out of vision. But the legend and map are bathed in the light of the moon; they take on the sense of a real place, in miniature; the shadows his hands make seem to increase its dimensions: there are shadows in the small indentations, like valleys; even a spot like crops seems to be growing on the carved wooden surface; the names of the places have faded, and now appear themselves as part of the natural landscape. He is praying; even his hands are sweating. And his sweat becomes as drops of blood running down to his wrists and into the valleys on the map; they turn into red rivers; they rise, and then they spill, and dance for a moment among the moonbeams, and then they drop in a pool on the ground, mingling with the blood that sweats from his face and legs. He kneels in a pool of blood. The blood sparkles and rills in the light of the moon. "Father," he says, "if this cup...," and with these words he produces a cup from under his garment — it is fashioned with gold leaf — and he holds the cup before him, thrust at arm's length, into the middle of the shaft the moon makes. And then he releases the cup; he opens his hand, and the cup stands, beyond his extended fingers, floats in the middle of the shaft of light: shimmers of gold leaf in the middle of a cylinder of

100

moonbeams, that seem to be constantly moving, bouncing like tiny diamonds, that flash with occasional blue light against the chalice. "Father," he says, "if thou art willing, remove this cup from me." And then, from a corner of his eye, he catches a movement in the trees about twenty feet to his left. And when he can see through the dusk, he finds it is Peter again, this time standing beside a tree in the garden; this time Hound is with him. He shuffles; he is afraid to approach; he whispers, "Master?" And Jesus sighs and bids him approach. And Peter does so, slowly, always keeping his eye fixed on the cup, which remains suspended in moonbeams in the air. He comes close to Jesus; when he steps in the circle of light, in the pool of blood, drops float up, like phosphorus, illuminated, around his sandals. He never once takes his eyes from the cup, though he speaks to the Master, saying: "Master, is it not yet time?" He seems only too glad to go back to the others, when Jesus bids him to do so.

The cup still stands in the air, and Jesus is looking, as if at another person, into the sky through the trees. The cup begins to move toward the sky. The face of the Savior is tortured with sweat. The cup stops, just at the edge of the trees. Jesus speaks: "Thy will be done." The cup moves back toward his hand; it reaches his fingers. It is full of blood, that bubbles with moonbeams, like pink champagne. He takes it out of its place in the light; he drinks it. And as he swallows his eyes fill slowly with blood; they bubble; they glow like twin red planets set in his face. And then he lowers his head to his chest; his eyes shine down on the legend. He replaces the chalice under his garment.

✝

The trees   The chalice   The moonbeams   The legend
The dog   The dark gnarled trees   Garibaldis
The rill of the brooks   The moss   The winding path
The valleys below   Butterflies
An ear on a patch of moss   In a pool of blood
Peter   The Agony   . . . if this cup
Occasional blue light

A resting place   A cylinder of moonbeams
Sandals in blood.

☨

This time, when he has finished, he stands up heavily and starts toward them. He knows already that they are sleeping before he takes a step. And then, before he takes a step, he sees another coming toward him, the one who's been missing, who comes up to him.

And when Judas approaches him, he comes slightly sideways. He is short and heavy, awkward. His arms raise, automatic, in a gesture of welcome. "They are sleeping," he says, and kisses Jesus on the lips.

And with that kiss is the garden flooded with light. A soldier behind every tree, steps out, and the garden is seen in its true light. There are insulated wires, winding between stones, that connect to dim lights at the bottom of the pools. Some of the goldfish are made of wood. Shrubs have been moved to accommodate the charming wind of the path. The trees have been methodically pruned; the moss has been brought here from a distance. Reporters squat under hoods at their cameras. The sound of a train can be heard in the valley below. The howl of a dog. Peter cuts off the ear of a servant. The ear falls on a patch of moss. The cause of the rill of the waterfalls is electric. Only the carved legend, the map, the blood, and the cylinder of moonbeams are authentic. The cup has entered his breast. He opens his garment and says: "Judas, doest thou betray the Son of Man with a kiss?" To the left of his sternum the flesh is transparent. His heart rests in the bowl of the golden chalice, in the middle of his chest. It throbs; it is imbedded with jewels, blue diamonds and rubies; it is crowned with thorns; moonbeams play in a halo around it. The heart sweats; and the sweat turns to moonbeams and blood, and bubbles over the rim of the cup. He places his index finger on his heart. The lights dim in the garden. The ear rises from its bed of moss. It is suspended in the middle of the air. He closes his garment; the ear goes and connects itself to the head of the servant of the high priest. Then they close and arrest him.

And as they take him down the path, their lights and torches marking the way, they buffet him between them. Peter lingers at the back of the crowd; questioned, he betrays him three times. A cock crows in the night. Judas remains, and hangs himself from an olive tree; the tree bends under unnatural weight; his body sways and lurches in occasional blue light; moonbeams shimmer in sweat on his chest. The things the legend represents are growing on the carved wood. The ground drinks up the pool of sweated blood. The moon passes behind a cloud. The moonbeams flicker and die. Darkness closes on the garden forever.

THE SCOURGING AT THE PILLAR

The pillar itself is made out of salt, a spike drives up through its center. It is six feet high, a cylindrical shape; its head is rounded. At the other end, the spike drives into the ground. Eight inches thick, it stands up straight in the center of the room; at its base is a square salt block.

There are two metal rings, one on each side of the pillar; they have tied him to the pillar, and then they have gone away. He is alone for a few minutes in the empty room. The floor of the room is made out of hard-packed dirt; there is a large table, directly in front of the pillar; one small window, high in the wall to his left, lets in a little light. Then they come back again.

There are three of them. The first is large; his tunic is open at the throat; when he enters, bending through the low doorway, he rips out a tuft of hair from his chest. "Hail, King of the Jews," he says, and throws the tuft at Jesus. The second one enters, dragging a large wicker chest behind him. The chest contains a purple robe, a reed as long as a staff, a few whips of various size and description, a container of salt, eating utensils, a couple of bowls, clothing, and other things. He drags the chest up to the side of the table, sits down, and begins to rummage through its contents.

The third one is short and enters the door straight up. He carries the severed head of a dog on a silver platter. He too sits down, placing the platter so that it sits in the middle of the table; the dog's head faces the Master; it is the head of Hound. There is also a dwarf.

The dwarf enters and moves directly to the table. He carries a basket of thorns and vines. He pulls out a stool and sits down at one end of the table.

Hound's head faces the Savior; it is matted with blood; the hair is twisted in matted tufts; the lips are pulled back from the teeth to the gums, which are no longer pink, but now have the color of cooked liver; the nose is dry, the eyes clouded over. But even as the Savior looks at the dog's head, the eyes seem to begin to clear, and then it is as if Hound were looking into his

Master's eyes, ready to scamper off, after a stone or a stick. And then the one who dragged in the wicker chest raises a cleaver and splits the dog's head open. But only a bit: enough to expose the brain. "Let's eat," he says with a snarl, and dips his fingers into the dog's head.

And then the three bring out napkins and utensils and begin to eat of the head of the dog; and all the while they watch the face of Jesus. "Would you like some of this?" the big one says, holding a piece of brain on his fork. But he sees that the Master is looking intently into the eyes of the dog, that he is not watching the fork. "What's this?" the big one says, and spins the platter around on the table. He is faced by the clouded eyes; he plucks one out with his fork and pops it into this mouth. Then he turns the head back around. But Jesus keeps looking intently, and where the eye had been, and deep in the empty socket, another eye appears to the Savior, and he keeps looking into it. The large one sees this; he jumps up and runs around the table. But when he looks into the dog's face, he sees only the one-eyed look there. "Let's get on with this," he says. "What'll we do first?" The small one says, "Let's dress him up." All the while, the dwarf sits alone, weaving a crown of thorns.

And then they loosen his bonds and dress him in the purple cloak. They put a reed into his right hand and strike him in the head with another reed. "Hail, King of the Jews," they say, bending their knees; they do him homage, laughing and spitting upon him. "Let's make him lick the pillar," the large one says, and they make him do that, starting at the head and working his way to the base, until they are sure he is very thirsty, and then they offer him water, but pull the cup away from his mouth before he can drink.

But Jesus continues to stand straight up, and they are dimly aware that their mocking is not working to humiliate him. So the small one walks up to him and strikes him in the mouth with his closed fist.

Blood flows from the corner of the Savior's mouth, and his tongue comes out of the corner of his mouth to remove it. But, O, the sight of that tongue! The three are stunned by the look of it. It is red and the color of unimaginable wine; its texture is that of silk. It is as if they are seeing their own loved hearts be-fore their eyes. And though the tongue only comes out for a moment, it seems to them that the memory of its appearance is as real as the beautiful tongue itself, lasting for long agonizing moments before their eyes.

"That ought to be enough," the middle one says, and he goes out to tell the guard that they have finished. But the guard an-swers him in amazement: "Why, you just now walked in there! You've only been in there about ten seconds!" Embarrassed, the middle one returns and tells the others. The dwarf keeps work-ing at the crown of thorns.

Now the three put their heads together, and while they are talking Jesus gazes out through the small window to his left in the upper wall. The first thing he sees is a bit of cloth, high in the sky, which is falling, taking on the shape of a small cape as it flutters down toward the earth. And then, from behind a hill in the distance, he sees Hound rising in an aura of blue light, flying free, needing no cape, higher than the two of them have ever flown before, gracefully lifting up in the distance. The cape comes to a fluttering rest on the edge of a cloud bank; it appears to him as a mere dot of blue on the low clouds. Then the body of Hound enters the bottom of the cloud. And when he emerges from the top of it, and as if on a whimsy, he reach-es out with his teeth as he passes the cape and grasps it; he throws the cape with his head, spinning it onto his shoulders. And then the vision of the dog diminishes as he enters the high-er clouds, his cape rippling around his shoulders.

They decide to dress him up like a woman and to make him dance. There are women's clothes in the wicker chest, and they take them out and put them over his body. He is dressed now in a long blue gown that is cinctured in the Empire fashion, high up under the bodice; they put a blue veil, like a babushka, over his head. "Now dance for us, harlot," one of them says, and they push him and poke him before he can begin. And then he is humble before them, and he begins to dance.

They laugh at his copy of feminine movement, and when he circles around the pillar and comes close to the table at which they sit, they throw jibes at him, and spit upon him, and throw pieces of hair and flesh from the dog's head into his face.

But then his dance begins to take on a pattern. And when this happens, the bodice of his gown seems to fill up with substance. (They assume it is air caused by his movement, but it looks, uncomfortably, like flesh.)

The dance is a minuet, and the Savior dances as if he were moving with a partner. His gestures are exact and perfect. When he raises his hand in the air, it is as if there were another hand within it, the hand of the masculine partner, turning him, in a pirouette and a bow. And as he dances his gown spreads out from his feet; it takes on the shape of a full dress with a hoop; behind him seems to appear a bustle. The veil takes on the shape of a delicate mantilla, rising, shaping itself over his head.

And soon they are not laughing and spitting, but just sitting and watching. And then they begin to hear music, of a kind they have never heard before: stringed instruments, and brasses, a harpsichord and cello. They sit that way for an hour, transfixed by the Savior's motions; even the modest room seems to take on an elegance in their minds, and it is only a light cough from the dwarf that brings them back to themselves, each a little unnerved, a little abstracted.

"You pig!" the large one screams, and slashes at Jesus with his knife. "Give me some salt!" he yells, and begins to slash at the Master's arms, rubbing the salt into the wounds. "I'll teach you to dance! I'll give you something to dance about!" And tying him back to the pillar, he begins to kick the Master's shins and knees.

But though the Master groans at the pain, he remains constant; the expression on his face is still placid, his smile — neither a grimace of pain nor ironic comment — is a smile, unmistakably, of love.

"The pig," the large one grunts under his breath. "I'll show you what pain is." And then he places his own hand flat upon the table and severs his own index finger with the knife. He throws the severed finger at Jesus.

"You see how tough we are," says the small one. "Watch how I strike myself." He extends his hand out from his shoulder, forms a fist and strikes out at his own face. But while he is doing this, he is looking into the eyes of Jesus, and his hand stops a fraction of an inch before his nose. The next time he tries to strike himself, his head lurches back before the fist can strike him. And then he moves to the wall, and places the back of his head against it. This time he looks at a spot on the wall just to the left of the Savior's head, and when he strikes at himself he shatters the bones in his nose. "He sees now," he whistles in pain through his crushed face. "Now he shall shudder in fear." And the three surround Jesus with their knives and begin to move closer to him. "Ahem," the dwarf says, "remember, you must not kill him."

And the three slack off a bit, considering what they can do to humilitate him. They go to the wicker backet and look into it. "Whips?" the middle one asks. "What the hell good are whips?" says the large one. "We've already tried cuts and salt." "Wait," says the small one, "maybe it's time now." And he goes out and says to the guard, "Is it time yet?" But again the guard is amazed by the question. "Time yet? Why you've only been in there for twenty seconds!" And the man is embarrassed again, and goes back inside.

And then the three begin to mutilate themselves even more. They strike each other with whips; they cut pieces of flesh from their bodies and pour salt into their own wounds; they grasp their own genitals and twist them until they scream. And constantly they watch Jesus while they are doing these terrible things, but the Master expresses no hint of fear for them to see.

And then, after a while, Jesus takes pity upon them and utters his first words to them: "It's time," he says, and the small one rushes out to the guard and asks him. "Yes, it's almost time," the guard says. And the small one returns to the others and tells them: "It's almost time." "How about a little refreshment?" the dwarf says, and from under his cloak he produces a small flask of liquid. And the three drink from it, and they even offer some of it to Jesus, but he declines to take it, and the three look at him with worried looks on their faces. And he speaks to them, saying, "Do not be fearful. I shall hang my head as I go out; come now and spit upon me some more, that the humiliation shall be visible when I go out." And the three approach him, and gently they spit upon him, and each, as they come close to him, touches at the purple cloak they have put around him, and

the large one whispers to him and says, "Master, forgive me."
And Jesus answers him, saying, "It is nothing; I forgive you."
Then the three return to their seats, feeling suddenly exhausted,
and they each fall into a deep sleep; their heads gather around
the mutilated head of the dog on the table top.

"Well now," the dwarf says. "That was very good, very good
indeed. But it is not over, and what I have here will certainly
teach you something. For I am the devil, you know, and it will
be a pretty thing to consider that I, the devil, shall be crowning
the Savior of Man with a wreath of thorns!"

And then he produces the wreath; each thorn is exactly one
inch from its neighbor, and the length of each is exactly one
half inch. And he loosens the Savior's bonds and offers him
a seat at the base of the salt pillar. And Jesus sits down as the
devil requests, and then the dwarf places the crown upon the
Savior's head, pressing it into the scalp. But the thorns refuse
to penetrate, and the dwarf removes the crown. And then he
proceeds to separate the hair in a circular part on the Savior's
head, in order to allow a place for the thorns to enter the scalp.
Maybe the Master's hair was the problem. And then he again
places the crown on the Master's head, and placing the weight
of his small bent body upon it and whispering "Hail, King of
the Jews" into the Master's ear, he thrusts the crown down into
the Master's scalp. But again the thorns refuse to penetrate; the
head seems as hard as iron. And the dwarf climbs upon the Master's knees. He stands on his knees, the Master's head at his
waist, and pushes downward with all his might, in order to
force the crown onto the head of Jesus. And as he presses, the
crown gives a little; a few of the thorns buckle and bend away
from the perfect circle. "Pig!" the dwarf screeches, jumps down
from the Master's lap, and then goes back to the table and
straightens the thorns out. But as he is doing this he catches the
eye of the Master upon him, and he sees the irony of his smile.
And then in disgust the dwarf casts the crown into the air, but
it does not fall to the ground. Instead, the crown hovers and
floats in the air over the Master's head, and a blue aura of light
gathers around it. And then it gently descends onto the line of
the circular part, and the scalp opens to allow the thorns to en-

ter; and before the dwarf rushes out of the room, he sees the eight perfect lines of blood. They flow at exactly the same pace, down the forehead and across the placid face of Jesus.

THE AUTOGRAPH

The veil hangs in a shadow cast by the failing sun. There is a kind of light breeze enters the window, causes the veil to billow out, like a small sail, from where it is hung, a pin at each corner, on the wall across from the bed. Nobody's home. The sun sets. The innocent veil billows out in the light breeze; a pin is dislodged from one of its corners; it spins in the room's air, shining, then drops without sound, on the wood floor. The veil's free corner flaps with inaudible taps against the wall.

The moon is framed in the open window. The room is bathed in cool light. And now the other pins, strained loose by the gentle breeze, spin away from the veil and the wall. The veil falls, floats on the room's air, and then it settles, face up, on the floor at the foot of the bed.

✝

Veronica comes to herself with the veil in her hand. The sun had been hot, and she'd been standing at the back of the crowd, at a point halfway up the hill, waiting for Jesus to pass. A place in the open sun. She'd been late, and the places along the way that offered shade had all been taken. She'd worn the veil over her head, protection against the rays of the sun, but then a cloud had covered the sun, the place became suddenly cool; she'd taken the veil off. Jesus rounded a corner of the path below her, and when he reached the place where she stood, he fell for the second time.

✝

The veil rests on the floor at the foot of the bed, but the bedroom window is open. The night is bright and cool. A full moon. No sign of rain, but a light breeze enters the window again. The veil stirs, as if there were mice running under it. And then it rises slowly, and floats at a level with the bed. Against all the laws of physics, it floats to the open window. And then it leaves the house. It floats over the landscape, at an average height of four feet above the ground.

✝

One end of the crosspiece rests on the dusty ground. The other thrusts up into the air. Jesus is on his knee, the cross resting against the side of his body. A man stands behind him, whipping his calves and ankles. Two men are dragging him to his feet again. His face hangs down from his shoulders. He is bleeding profusely and sweating, and the two liquids are forming a pool on the ground under his face. And then he turns his face to the crowd, and immediately the crowd parts, and as he looks through the corridor of bodies, he sees Veronica standing, facing him, holding the veil in her hands.

And then Veronica stepped forward, holding the square veil by two of its corners, and Jesus looked into the veil. And then he released the cross for a moment, reached out, and took hold of the other two corners of the veil. And then he slowly thrust his face forward, his neck seeming to grow longer, and pressed his face into the center of the veil. And then the two men managed to drag him to his feet again, and he continued up the hill.

Now Veronica did not wait for the Crucifixion, but folded her veil and put it under her garments and went directly home. When she was safe in her house she removed the veil from her clothing, opened it, and attached it, with straight pins at each corner, to the wall facing her bed. And then, as the veil dried, she saw the face of Jesus upon it.

✝

The veil is pinned to the wall of Veronica's bedroom; it has hung there for two months. It faces the bed. Veronica has built a structure under the veil: a table covered with a cloth, two candles at the sides of the table; they frame the veil. She has placed a cushion on the floor in front of her altar. She prays, daily, before the veil.

In the beginning she had brought in her neighbors to look at the veil, but some of them had laughed at her for making the altar, and now she only allows her boyfriends into her bedroom, insisting they pray with her at the altar. Only after they have done this will she sleep with them. But when she is alone in her bedroom, she gazes for hours at the face of Jesus.

The face of Jesus is perfectly symmetrical, and this is what gives it its energy. The features are perfectly set in the face. Each hair of the beard on the left side is mirrored by a hair on the right. The face is totally without motion. Veronica has held a mirror, sideways, up to the middle of the face of Jesus. She has done this for hours, counting each hair of the eyelash, then counting the mirror image of it, then counting the hairs on the other, real eyelash. There is no flaw to be found in the symmetry.

Faces move, have asymmetrical features. But the face of Jesus is perfect, kinetic energy, always on the edge of potential movement. The eyes look straight out of the face, but wherever Veronica stands in the room, the eyes of Jesus are fixed upon her.

✝

The veil floats over the moonlit innocent landscape like a veil of tears. It covers the meadows and hills, following the contour of the land. The intractable image of the face of Jesus looks up from the surface of the rippling veil, into the starry night. The sky is so clear that each visible star is distinct in its relative brightness, and each of a different hue. There are auburn stars, stars like various shades of coral, stars the color of cut limes, light blue stars; some are vermilion. The eyes in the face of the image of Jesus are clear deep blue; they are completely dilated; they have no pupils. When the veil passes under the moon, the eyes of Jesus contract, change to a hazel color, then back to blue, and the veil passes from under the moon's light.

The night is completely clear; only the moon and the stars are visible. But then, from over a hill at the edge of the landscape, a black cloud appears. It moves across the sky; it passes over the face of the moon and comes to rest.

The moon is no longer visible; the stars have faded into similitude. The veil ripples on the darkened air, and then it slowly rotates, turning the face of Jesus toward the ground.

113

Veronica rolls on the bed with a young man. She is twice his age; he awkwardly fondles her large breasts; she guides his innocent hands over her body. The room is only lit by the candles, which flicker and cause the illusion of frantically jerking bodies. The young man is impatient. Veronica hauls him over, onto her belly; she spreads her legs, guides him into her body. He buries his face in her neck. Over his left shoulder, she catches a glimpse of the veil. The young man's head pulls up from her neck; he begins to kiss her face. She thrusts his head back into her neck. Now she can see the veil again. The eyes of Jesus are upon her; the face seems sad and pained. On the left cheek, running down from the eye and into the beard on that side, are evenly spaced, alternative drops, of blood and tears. Veronica throws the young man off her body. He barely has time to gather his clothing, before she rushes him out of the room.

†

A farmer is unable to sleep. He dresses in the night, proceeds to go out to the barn to check his stock. On the way he looks up at the sky; he is amazed by the moon, by the various stars. Standing in the middle of the barn he is struck by the beautiful objects he owns. The halters, saddles, and wagons, even the stalks of grain: all seem strangely beautiful to him. His one horse stands in the stall. The farmer looks into the horse's sleepy eyes; for the first time he thinks of the horse as his partner, he marvels at the beauty of the horse's spine, that gentle slope along the smooth broad back. The horse raises and lowers his head; he gives a quiet snort. The farmer turns toward the open door. Over the fence at the edge of his yard he sees the veil passing, rippling, slowly over the innocent landscape. In the morning he tells his friends about his vision. No one, of course, believes him.

†

The veil seems now intent on its passage; it heads on an even line toward the sea. But it goes slowly, and at times it diverts from its path: it stops, and gathers around a fence post, like a paper stopped in its flight in the wind. Sometimes it enters into barns and houses. It rests for a moment on a kitchen table; it enters the open door of a barn, hovers for a moment in the air, then rests like a saddle blanket, on the back of a sleeping horse. Often it brushes the tops of the growing crops, gathering seeds, that cover the face of Jesus, and then proceeds for a while, and then gently shakes the seed into the fertile soil of another field.

On stones, on the handles of plows, spread out over closed windows, covering the tops of fence posts, gathering seeds, brushing the wings of hovering nighthawks, floating on the surface of still irrigation pools, tapping against the sides of barns, resting, sometimes, flat on the ground, then rising; the veil continues, over the changing landscape; the air becomes salty and damp; the veil reaches the edge of the sea.

✝

He remembers his mother working in paisleys and needle point. The early years of his life, at home; she'd sit in a chair by the window, sewing at shawls. The light would often come in, causing the needles to sparkle and flash in her hands. In reality she was very slow, but careful, and he would sit for hours, watching her moving hands in her lap. Once she cut off a small swatch of fabric and gave it to him as a bookmark. And a veil she had made, she'd given to him for a scarf, and the few trinkets his father had given to him: a small carved figure of a camel, a wooden ring, an old piece of leather halter to use as a headband.

And he had taken some tools and wood, and built a small box to keep his possessions in. The box was awkward, crooked, the wood too thick; the lid didn't fit. But he had kept it under his bed, and often at night he would take it out and handle the objects: a necklace his mother had made out of olive pits, a few dried leaves, a piece of a broken fingernail; he'd put the objects in the center of the veil, tie up the four corners, put the veil into the box, replace the box under the bed, and sleep above it.

†

Veronica sits, propped up on the pillows of her bed, watching the face of Jesus. The veil is stretched tight over the wall. The candles have burned down and gone out, but the face on the veil is bathed with an inner light of its own. The deep blue eyes of Jesus are upon her, but the rest of the face is incredibly peaceful, without expression at all. The face shines out of an oval of light defined by its structure. The thorns in the crown are perfectly spaced, and the holes where the thorns enter the scalp of the Savior are perfect also, not ragged, but exquisite receptacles, holding the thorns in place. Each curl of hair that falls over the forehead is balanced by an exactly identical curl on the opposite side; the short beard is exactly rounded. The face looks out from the exact center of the veil, which is two feet square. The veil surrounding the face is of light gray in color. The weave of the fabric is very tight.

Veronica does not pray, but simply allows her eyes to roam over the face of Jesus, learning its every line and hair, studying the lovely shape of the nose, the quiet line of the eyebrows. And then as she watches, she sees that the pupils of Jesus' eyes are changing in color, from black into red. The pupils widen, and as they change in color, they begin to spin, slowly at first, then faster, like small red wheels, small burning suns, spinning in the eyes of the Savior.

Then, as she watches, she sees that the side of the left cheek is changing, the line of the cheek gyrating, then breaking apart into small lines and fragments. And a tuft of hair from the side of the beard begins to wave, like a miniature field of high crops in the wind. The hair collapses, drifts from the side of the beard, joins with the broken lines and fragments of cheek, falls from the face onto the side of the veil, and begins to form itself into letters. And then the letters take shape into words, and as the words form a drop of blood bubbles up like Burgundy wine from the side of one of the holes holding a thorn. The drop extends itself like a pear-shaped drop of water, but when it falls from the thorn it gathers into a red transparent ball and floats in the air, reaching the extreme left of the veil, and hovers above the words of the now completed legend. The ball, like a small red crystal globe, hovers above each word, as Veronica reads, from right to left: *This Is The Story Of My City.* As the ball passes from each word to the next, it glows brighter for a moment, it pulses, and the word dissolves again into lines and fragments, and returns and reforms, into the face of Jesus. When Veronica has finished reading, the ball remains, pulsing, on the light gray open field of the left side of the veil.

And then, after a moment, the ball begins to move again. It circles the face; it hovers over the head of the Savior; it pulses more strongly, begins to change, into a tongue of fire. And the hair on the top of the Savior's head, contained in the ring of the crown of thorns, begins to wave and break into lines and fragments, and the fragments rise above the head, and form themselves into a rough-hewn house, and the tongue of flame lowers as the house rises over the head of Jesus, and when the fire touches the roof of the house, the boards of the small structure gather themselves into perfection. Veronica sees a small, tightly constructed house flooded with light through its open windows, and the wooden door of the house changes to gold, two high double doors, intricately carved, magnificent, and the doors open slowly upon a chamber lined in red velvet. And in the middle of the chamber is a tiny chalice; but she can see every sworl and stamp of the fine-wrought precious metal. She sees that a small heart, like the heart of a dove, rests in the chalice, floating in blood. The heart is crowned with tiny thorns; it is imbedded with occasional gems of many colors. And as she watches in wonder, the tongue of fire forms once again into the crystal ball of blood, and the small house dissolves, and again becomes the hair on the top of the head of Jesus, and the face returns to stasis.

And all night long Veronica watches and wonders. Some things appear on the veil that she recognizes: a meadow filled with ripe olive trees, the olives falling like slow rain to the ground; an old mangy dog appears, walks to the edge of the veil and dissolves, becoming again a tooth in the mouth of Jesus. But there are wonders she has never seen: incredible weapons and machinery, carts that move without being pulled

117

by horses, in which people sit and ride, with veils tied round their necks, blowing out in the wind behind them. And many messages also, and groups of words: *I Am The Way, The Truth And The Light; The Meek Shall Inherit The Earth; Bless This House; The Prodigal Son; On This Rock; Realized Eschatology.*

All night long. And then in the early morning hours, but before the daylight is present, the crystal ball of blood returns to its source. It floats up, gathers, and folds around the thorn and slides back into the thorn's receptacle, the hole in the scalp of Jesus. And then, as Veronica watches in wonder, the face of Jesus begins to crack down the middle, a perfect line, from a thorn in the middle of his forehead, directly down the middle of his nose, between his two front teeth and across the beard on his chin. And when they have parted, the two halves of the face turn slightly inward, toward each other, but not into complete profile. Both of the eyes of Jesus are still upon her, but each glows red from out of each of the two identical halves of the face of Jesus. And then the two halves, slightly in profile, are beheld by Veronica; and then she realizes they are two complete faces; it is like when she had held the mirror up to the center of the Savior's face. The two halves are identical; the two missing halves are, because of this, unnecessary. She realizes she is running her hands over her body, and immediately she knows there is nothing impure in this. She is touching her self in wonder, for the first time in her life, her hands like the hands of her various lovers, but not like that at all; because they are her own hands, she knows them intimately, they are gentle and kind to her body, they understand her.

And then, as Veronica watches the twofold face of Jesus, the bright red eyes of the Savior leave her body; they rotate slowly, and then each eye is looking into the eye in the other half of the Savior's face; the two halves move closer together; the half lips touch; the half tongues entwine; the cheeks are together; the faces merge into one again. And for a moment the eyes rest upon Veronica again, and then they begin to roll, back into the Savior's head; they rotate completely; the backs of the eyes are milky white. Veronica sees the legend appear, in exquisite blue letters, one word embossed in the milky globe of each eye. The words appear in her own dilated pupils; they read: *Know Thy-self.*

✝

He thinks he could know his mother more intimately, taking the objects out of the box, opening the veil on the coverlet of his bed, handling his possessions. But his mother remains an enigma to him; there is a certain knowledge he cannot attain. And yet he's aware that the knowledge is possible, precisely because he cannot attain it. He places the objects back in the box. He sleeps over them.

But most of the objects come from his father, but he no longer thinks of them as masculine. When his father had given him the various pieces of wood he had noticed the lines of the knife in them, had smelled the horse sweat in the headband. But putting the objects beside the ones from his mother, wrapping them in the veil, they had all seemed to soften, to become like his mother: a possible faint scent of perfume from the veil, a delicate new feel to the small wooden camel. "How does this camel feel to you?" he had asked his mother. But she, though she loved her son, was somewhat afraid of him. He said strange things often, and she was guarded with him. "Oh, nothing special, I guess; like a small wooden camel," she had said, and though he'd looked intently into her eyes as she spoke, he could find nothing there that he understood.

The day that she gave him the veil is important, he thinks, emblematic. He was playing with Hound in the corner of the room. But his eyes were watching his mother. She had the veil spread over her lap; it was full of olives, and she was carefully removing the stems that pulled loose from the tree when the olives were picked. She placed the cleaned olives in a wooden bowl, on the floor, beside her chair.

And when she had removed the last olive from the veil in her lap (an indentation still remained where the weight of the olives had rested) and had pulled the small green stem out of the olive, placing the clean olive in the wooden bowl, he saw her lean back in her chair, and just then he had run a splinter into his hand as he pulled it along the rough floor.

He remembers leaping up in his pain and running to his mother's side, but he had not cried out, and his mother was still reclining in the chair, her eyes closed, her arms hanging down at her sides, her head on the wood of the chair's back, the veil resting between her slightly opened knees. And though he was having great pain, he stood for a moment in front of his mother's spent body, somehow afraid to disturb her rest, and then he fell to his knees before her, and buried his face in the veil, and began to weep. And then his mother placed her hand on the back of his head, pressing his face into the veil in her lap. And his nostrils were flooded with a strange mixture of scents, of olives, fabric, and strange perfume. And before he was really sufficiently soothed, he removed his tear-stained face from the veil and got to his feet. And his mother smiled at him. And then she handed him the veil, saying: "Here, take this beautiful veil now; you can keep it for your own. Wipe your eyes."

†

The face of Jesus rests in a bed of grass. The veil ripples slightly out from where it is moored, one corner pinned to the ground with a stone. A slight breeze. A promontory edge of the sea. The right cheek is pillowed in grass in a slight concavity in the ground. The right eye is closed. But the left eye of Jesus can see the beginning of day; first hint of the sun shines through the veil to his left. The veil billows and ripples slightly. The eye seems like a man with a sheet pulled over his head, as if in a tent in the morning, from under the changing domed space of the stirring veil. And then the veil pulls loose from its mooring and rises six inches off the ground.

And now the veil extends itself and becomes stiff; its wrinkles depart. It hangs like a thin piece of tin, suspended from invisible wires, their source somewhere in the empty sky, on a cliff's edge, a half mile above the sea. And then the veil tilts, its sea edge downward, and then its downward corners begin to slowly fold back; they meet in the center of the veil, creating a single needle point of tin as a downward edge. Then it begins to move, gradually at first, and then it is flying through empty air, attaining terrific speed. It reaches the viscous surface; its point enters into a wave; it makes no splash, leaves no immediate evidence of its entrance.

The gulls come back and sit on the sea's surface, where it is mostly placid. But under the sea the veil is moving, now slack again, like a small ray, a possible swimming octopus. And as it goes it occasionally rests on the backs of fish: the face of Jesus pressed into the skins of sharks; the curve of his lips imprinting their seal on the snakelike bodies of eels; the lines of his beard mingling with those on the etched surface of the shells of turtles. And shellfish, the expanses of coral reefs, the bodies of whales, the veil formed into a tube, around a bouquet of sea grasses. And once the veil gathers forty oysters: it forms above the bottom, in the shallow water over the bed, like a small submerged falling parachute; it gathers the oysters into its dome; each oyster touches the face of Jesus, and the parts of the face leave their autograph upon it.

Time passes. But for the veil, that floats in another element, there is no time, only continuous present. For the future is certain and no matter of striving. The veil moves in a leisurely way; the face of the veil doesn't age, nor is it worn away, like the miniature crucified Christ, held in the priest's hands at the altar, the foot now smoothed and undefined, from too many lips brushing against it. And the big fish eat the little fish; and gulls fly to high altitudes, drop clams on hard rocks to open them. And each animal that eats another takes a part of that animal being into itself. And the face on the veil has left its autograph on small creatures. And large creatures take a part of that autograph upon themselves. It appears, visible, on the various shapes and textures of body. And people drop fish nets into the sea. They ride out to sea in charter boats, hanging a multitude of lines over the side. And many old ladies walk on the beaches, searching, collecting shells with incredible vengeance, and are thought to be strange and eccentric; the shells hang, drilled out and strung on pieces of leather as room dividers; they turn and flash their intricate lines and sworls, when they are brushed against, when sun enters the window.

And the veil moves leisurely under the seas, through the

channels and waterways of the earth. It waits for a time when it can emerge, can go and hang in a prominent public building, its replica fixed to the sides of large structures, in warehouses, in homes, in public toilets. And each man and woman who steps before the veil will see a familiar face. It will be his own face hanging before him, in the time of the Realized Eschatology.

†

Veronica walks down the path from her house on the way to the Crucifixion. The sky is clear, and the hot sun beats on her head as she goes, out in the open, away from the cover of trees; she pulls her shawl up over her head to protect herself from the sun.

But the sun is hot, her shawl of a wide weave, and by the time she gets to the main road that leads to the town she is dizzy with the heat.

The road is full, with country people going to town for the Crucifixion, and because she is dizzy, unsure of foot, they bump into her, jostle her in their haste, and she is continually buffeted among them. Halfway to town she is tired and decides to rest by the roadside, to wait for the bulk of the people to pass. She sits by a bush at the roadside, watching the crowd thin down to a trickle. She wipes her brow with her hand, reaches down to wipe her hand on the bush, and her hand comes to rest on a piece of cloth, a thin veil, tangled in the branches. And when she looks up from the veil in her hand, the road is empty. She places the veil over her head, ties it under her chin, gets up, and continues walking, the veil protecting her head from the sun.

†

And until the time had come for him to begin his public life, he had kept the box of belongings under his bed. He examined them nightly, spreading the veil over the coverlet, studying each object, running his hands over them, touching them against his face. And then the time had come for him to leave, and for

122

the last time, the night before his departure, he had taken the box out, opened the veil, and looked upon his objects.

And then he had taken each object and placed it into his mouth and eaten it. And in so doing he had taken the objects into his City: the headband hung on a nail in an empty house; the camel was tethered to a hitching post in his City. Even the bit of fingernail, which he chewed up and swallowed, became whole again as it entered his City, and came to rest on the floor of a stable on the outskirts.

But he could not bring himself to deal with the veil, and he decided it would not be an encumbrance to him. And so in the morning, before his leaving, he had placed the veil round his neck as a scarf, had tied a knot in the front, and had left with the veil as part of his clothing.

As the years passed he found he had use for the veil, and he wore it always. Sometimes, when the days were hot, and he had traveled a long way in his preaching, he would rest for a while at the side of the road, and when he removed his veil from his neck and wiped his sweaty brow, he would feel a pleasure of coolness against his temples, would notice a vague and distant scent in his nostrils. And then too, he would use the veil often as costume in his preaching. Relating some parable, he would find himself altering his voice to conform to the difference in characters, and when a woman appeared in a parable, he would often place the veil over his head, as a babushka, and speak in a woman's voice. And when he did this no one laughed at him, for he was very good at pretending he was someone else, and the veil played no small part of help in these matters. When people gave him bits of food to eat, he would always remove the veil, spread it over the table or ground, and place his meal upon it. The slight formality always pleased him, giving his meals, no matter how meager, a certain sense of occasion.

But no one ever took particular notice of his clothing; no stories grew up about the cape that he wore over his shoulders; no one ever mentioned the veil. But the cape had been made by his mother and the veil had been given to him by her, and they were precious to him.

And when they had finished the Last Supper, he asked to be

left alone for a moment, and his apostles left him alone in the narrow room; and Hound was beside him. And he sat stroking the back of the dog, whose head was raised, who was looking into the eyes of his Master. And Jesus took the veil from his neck and placed it for the last time over his head, and tied it under his chin. And then he spoke to Hound in the voice of a delicate woman. He told Hound the story of Noah; he spoke of the Hound of Heaven; he prophesied the coming of St. Francis. And then he removed the veil and tied it around Hound's neck, and then he spoke to the dog in his own voice. "Go and hide yourself," he said, "for you are known to them. But remember whatever happens, you shall sit by my side in heaven." And Hound pressed his muzzle into the Master's palm, and he closed his eyes for a moment and licked the palm. And then Hound trotted to the door, his cape rippling across his back, the veil tied like a collar around his hairy neck. And when he had reached the door, he turned for a moment, and then he lifted his paw in the air and bowed his head to Jesus, and then he ran down the stairs, the door slowly closing behind him.

✝

The farmer gets up in the morning and makes his breakfast. It is not yet light, but as he is cooking his cereal, on the stove in front of the window, he sees the various objects come to distinction. First the posts of the fence and then the wires, stretching between them. He is struck by the way the objects form out of the darkness as if they were coming into literal existence as he stands and watches. In no time at all, it seems, the various crops assume their specific identities: the yard, the flowers, a patch of newly turned earth; a birdhouse, and then the limb on which it hangs, and then the entire tree. And he feels as if he is, in a sense, seeing these things for the first time, or at least in a new way. And then he shakes his head and chuckles to himself, and then he remembers that he'd risen in the night, and then he remembers the veil.

Immediately the farmer removes the pot from the stove; he hurriedly pulls on his garments. And then he rushes out to his barn. He is six feet out of the door before he lurches, amazed, to a stop, for he is standing in utter darkness. He backs up then, enters the door again, looks through the window; he sees the barn bathed in the morning sun. And then with a shaking hand he prepares and eats his breakfast.

And when he has finished and ordered the room, the bed made up, the dishes restored to their places, he dresses himself in his finest garments; he washes his face and hands thoroughly, and then he goes out to the barn.

He places the bridle upon his horse; he prepares the saddle and blanket. And then he proceeds to curry the horse, and when he begins to brush the horse's back, he looks down into the hair and discovers the face of Jesus. And though he is amazed at what he sees, he is not fearful. He pulls a stool up by the horse's side, for he is a short man; and then he climbs up on the stool; he looks directly down, into the fact of the Savior.

The face is shaped in the hair, a matter of configuration, and when he tentatively brushes over the face, the hair comes straight for a moment, but then it swirls, returning to the natural waves and curls on the horse's back, and the face of Jesus comes into focus again.

And then the farmer, a timid and lonely man, straightens up on the stool and stands for the first time in his life at the full extent of his stature. He feels a laugh begin in his abdomen, and then, rejecting the use of a saddle and blanket, he throws his leg over the horse's back, sits down on the animal's spine, grabs up the reins, and digs his heels into the horse's flanks.

And the horse races, magnificently, down the road in the morning light towards town; the farmer is fixed to his back. And the farmer sings in the morning air, in a language never heard on this earth, and he feels an increasing fire grow up in his loins. And before noon, the farmer returns by the same road, his new and beautiful wife astraddle the horse's body, in front of the farmer and facing him, her buttocks split by the spine of the horse's back.

✝

One of them holds the dog by his tail, and then the tail goes slack in his hands, and a fountain of blood rises from the dog's haunches, for only a moment, and then the muscles contract, and the blood runs in rivers down Hound's legs.

They have thrust a sword through his body, then rammed the sword into a tree. Hound writhes on the sword, and as he yelps and growls, a bloody froth rises, spilling over his teeth. And then the one who had held the tail is angry, for when the tail came loose he'd fallen into the bushes, so he takes out his knife and slowly castrates the dog.

One man is approaching Hound from the front with a club. "Watch out for the eyes!" another man yells, and the man with the club strikes Hound on the muzzle, breaking his jaw. Then they eviscerate him, and then they thrust his bloody bowels into his broken face.

And as he is dying, one of the men raises his sword over his head. He swings it through the air, and slices Hound's head off at the shoulders.

And as the head leaves the body, the veil falls, strangely unmarred by the dog's blood, into a bush at the roadside.

✝

And now is he coming up the hill, bearing the cross on his shoulder. The path is lined to the top with people; there are larger clumps of them, where there are trees for shade. He thinks: "It is like a parade, and I am the only attraction, a kind of float." From time to time he can catch a glimpse of the top, a long way off; they have already tied the two thieves, with rough ropes, to the flanking crosses; he feels they want him to hurry.

And then he falls for the first time. And after they drag him to his feet, he struggles around a corner in the path, and then he stumbles and falls again. And while he is on his knees, he glances into the legs of the people standing by the road side, and what he sees amazes him. There, through the forest of legs, blurred by the blood sweating out of his eyes, he sees the corner of a veil, stirring slightly, brushing against the legs of a woman. He knows, immediately, that the veil is his own.

And so he forces the crowd to part, the woman to stay in her place at the back of the group, and when this happens he sees the woman standing, the veil hanging from one of its corners along her leg.

And then, though they drag him up to his feet, though a man stands behind him whipping his calves and ankles, though blood bubbles from the thorn holes in his head, a complete sense of his past life rushes in upon him. It is not the past of his glorious public life, but the more distant, endearing, real past of his early years. For a moment he thinks himself back in his father's house. It is midday. His mother sits in a chair by the window; she is removing the small green stems from olives.

And then he sees, through the blood in his eyes, that the veil is opened, that it faces him and is moving toward him. And when it comes but a foot from his face, he releases the cross, reaches out, takes hold of the two bottom corners of the veil and pulls it taut in the air. For a moment he wonders: what does the veil contain? what's on the other side of it? what does that sweet scent mean? And then he extends his neck; he presses his face into the sweet cool fabric; he loses himself for eternity, as if in pressing his face in the veil, the veil had formed a cocoon and covered his entire body, soothing each place on his naked skin, guarding him in oblivion.

And then he removes his face from the veil, releases his hold on its corners; and as he does this, the total reality of time returns to him. He hears the deafening mixture of voices around him; he feels the whip biting into his calves and ankles; he feels the dust choking into his throat; he's aware of the blood in his eyes, the crown, the feel and weight of the cross, the two men jerking him to his feet. And even the veil itself: as he releases it, he sees his own face in static and utter repose for a moment, a fleeting vague possibility of being completely human, a possible answer. But the veil falls from his fingers; the face in the veil comes alive, rippling in the veil's folds in the hot quiet breeze. And his own face contorts with the strength he needs, called down from his father's house. He gathers his City around him. He stands up under the weight of the cross again. And then, with a strange formal dignity, not of this earth, he turns from the crowd in a slow perfect arc, his weight balanced and

held firm on his ruined ankles. He moves away from them.
Veronica gathers up the veil, and places it under her garments.

JESUS MEETS HIS MOTHER — his own voice

Now have I spoken to thee, father, enough (but one more time in the near future) but, this piece of a tree, this pegged wood, this spaced stone, on which I am made to walk in the turning street; it turns to ground at the top, but I am unable to see it.

Give me a heart for a chalice. Give me clean feet. Give me a new hat. A man is beating my calves and ankles; a woman stands in the crowd with a veil. I can see my entire past and future. Good-by.

[It is here that he falls for the first time. The woman with the veil starts forward, hesitates as he rises, and then moves back in the crowd. A man keeps whipping his calves and ankles; a man with a spear helps him to rise; adjusting the cross on his shoulder, he sees his mother standing in the crowd. *O Jesus, Your painful fall under the Cross and Your quick rise teach me to repent and rise instantly should I ever be forgetful of Your love and commit a mortal sin. Make me strong enough to conquer all my wicked passions.*]

She stands at the front of the crowd, at a bend in the road, and I can see her, tears under the veil of my sweat, and blood, dripping into my beard, and she approaches. "Mother" (but not spoken); the man keeps whipping my calves and ankles; they allow me to pause for a moment. She genuflects at my feet.

Mother, your glazed blue eyes, your faint halo, your falling blue garment, small breasts I can see now, as I lean under the weight of my cross. They hang in the domed space of your loosened gown. Spread slightly, pink-tipped, virginal; between them your navel is visible, the beginning of hair in your crotch, dark place in the cup formed by the arch of your ileum. And your arms, as you kneel, your hands raised to the sides of my face, fingers touching even my crown. And that I hold you, my left hand at your tricep, the other holding the cross on my shoulder from falling.

[*O, Jesus, Your afflicted Mother was resigned to Your Passion because she is my Mother also, and wants to see me live and die as a child of God. Grant me a tender love for You and Your holy Mother.*]

And your eyes, your eyes are still vacant and empty as if they saw through me. But now I can see into your body, through your eyes like a mouth, into the turning recesses of your bowels. And light breaks as my vision exits your anus, and I can see your right heel and ankle as you kneel before me. And the heel is soiled and calloused from too much walking.

Mother, if at this last point I cannot see you, but through you, can only see the shape of your sexual parts, which are terribly virginal, but not in a warped way; you are still fifteen years old, you are still waiting for the son to follow the father, for the son to give shape to the father's life, for the son to take up the cross of the father's existence. This is what I want to tell you:

You have not aged. You have not known your husband. You have given unique birth. Your breasts are beautiful. A halo encloses your head and your life. Mother, I can only speak in parables.

But I can see my entire past and future, a manger in which I am born and rest in your arms as if in the arms of a natural mother. Mother, your dry breasts: you put me under a goat in the stable to suckle, and watched me, and held the breasts of the goat in your hands. And in my future they have already written of this, given it names, have spoken with various ecstasies, about mothers.

Give it a place then . . . babushka hiding her face. She leans over a cradle or a casket. She weeps, she laughs; there are touching stories; there are jokes also.

Mother, in this tableau, as you bend before me on one knee, your arms raised to my face, our heads close together, our halos merging, my eyes gazing down at your breasts; how can I clearly say that all women are my mothers, and that they are also my possible lovers, if that were possible, and that in my City there is only one mansion, in which we can all live, and you can be the son when you want to, or a lover, or even a father. And if I wish, I can be the mother, can grow breasts and suckle you. And there will be such joy in this mansion in my City, that we will make a room, as a joke, and call it the *Room of Mothers*. And over the door of that room, we will make and hang up a sign. And the sign will read: *This Room Is Empty*.

130

THREE VIEWS OF THE CROSS

1 — supine

And bearing the cross for himself, he went forth to the place called the Skull, in Aramaic, Golgotha. Pilate wrote an inscription and had it put on the cross. And there was written, "Jesus of Nazareth, the King of the Jews," in Hebrew, in Greek, and in Latin. The chief priests of the Jews said therefore to Pilate, "Do not write 'The King of the Jews,' but 'He *said*, I am the King of the Jews.'" But Pilate answered "What I have written, I have written."

They have taken the cross from his shoulders; four men, in uniform; they hold the cross by its four points, three feet in the air, between them. Carried like a casket, they trot ten feet from where the Savior stands, and with practiced precision they lower the cross, together, and place it upon the ground. And then they have taken Jesus, now with some semblance of tenderness, led him as if back from a window to a sickbed, and have helped him to lower his body down upon the cross.

In this view he is supine; he lies upon the cross in the slight concavity of the hill's crest, and they are preparing to nail him in. He rests on his back; his two arms are spread, at right angles from the shoulders, his palms up; his head rests below the inscription; his left leg lies in the dirt along the cross; his right leg is bent at the knee; the foot rests on a block of wood, like a pedal; the right leg forms a triangle in the air; the cross itself is the triangle's base. A soldier stands by his right hand; he takes forth, from his tunic, a bundle of instruments wrapped in an oiled cloth; he opens the two-foot cloth, spreading its four corners, and handles and examines the objects it contains: three thick nails and a heavy steel mallet.

He is supine; his head rests, balanced on the hard pillow of his crown; his head turns; he looks along the line of his right arm to where the soldier stands, handles his instruments; light sparks from the hammer. He taps along the length of his nails, testing their tensile strength. The crowd shudders at the sound, bows deeper; the heads of the women duck under their veils;

they move back inches: what if he misses? What if the hammer strikes into fingers or wrist, the whole bones crushed?

But he looks then, straight up from the cross, into the void blue dome of the sky, the empty inverted chalice of the world; there is no single configuration: empty, of clouds, of birds, of moving air, of any sign. Cool blood runs beside his eye, droplets of cool rest, in the right corner of his mouth. Those in the crowd still watching, shudder at the sight of the pink arrow of his tongue, that wets his lips, with blood.

"Father, I have come to the end of my rope; I am devoid even of humor. I am sick of my body, of its bondage. I want emptiness. Take me; do not confuse me, be not a human father."

He looks up. And then he realizes or he remembers that he has never seen such emptiness, such clarity, from this position; remembers, only vaguely, the convex world seen from another location; the cup turned over, emptied of its contents, transparent, a cup turned and pressed upon a globe; the objects of houses and people, moving. Himself now, on the globe itself, the clear blue emptiness of the sky, his last look, banal, from the human element.

And then he becomes totally human. The thorns bite painfully into his scalp; he fears to look at his right hand; his body shudders upon the supine cross. And the voices of the people come to him, their weeping and murmurs; and the thief on his right pleads with him; the curses of the thief on his left; top of the hammer along the nail's length. And in defense, against such painful immediacy, he thinks of his human father; his humor returns to him: the irony of carpentry, the soldier, now kneeling, at his right hand. And then he is able to fantasize, to imagine that the sky's dome is his halo, that the whole world spreads out from him. He becomes center and alone; finally, he is totally human: only the empty blue sky, him lying supine beneath it. And then he is able to displace himself. The hammer, an object of exactly equal importance, enters the empty field of his vision, glimmers, very hard and heavy, at the end of its wooden handle. And as it is lowered, his eyes turn to follow its arc; he extends his fingers, presses his palm into the point of the nail. And as the nail enters his flesh, he gathers his fingers around its cool length.

132

And after they have nailed his feet and hands to the cross, they lift it, dragging its foot along the ground. And then they stand it up, like a ladder in air, and pack the dirt around its base. And then he can look at the people.

2 — The Indulgence

He stands in the crowd watching, his cape pulled over his head like a veil worn by a woman. He is tall, and the people in the crowd jostle him; those behind him push him, struggling for a better view. He moves aside for them when he is able.

The figure hangs limp on the central cross. The body of the figure looks like ivory, drained of its blood; the eyes are beginning to roll back in the figure's head, revealing their ivory whites. The shoulders of the figure are dislocated, the arms longer now. Only occasionally does the chest rise, but mostly the air only passes out from between the lips.

He stands watching, under the bright sun with the people, and then he notices the sky darken, imperceptibly at first; not clouds, but the sun itself getting darker, the light fading. And then the figure on the cross elevates its head slightly, some words pass from its lips; the crowd murmurs, and then the head of the figure slumps on the chest. And when that happens a sudden wind rises, and the sky is thrust into darkness, a sudden storm; the sun shudders and winks out, and the day becomes like night. And the people look up fearfully, into the black sky, guarding their bodies with their veils and robes, against the driving wind.

And while he is watching himself hang on the cross, and the people's heads are turned away for a moment, he accepts the Indulgence. Time stops around him; the people are fixed in their various postures and gestures; the wind stops blowing, and the figure on the cross dissolves, takes on new shape in the air, floats down, and falls on his body in the form of a new garment.

And as he looks at the empty wood of the cross, he sees the green shoots begin to sprout from it; it becomes like a tree, and the shoots form into branches, and buds form and then begin to flower. And then the whole cross becomes a tangle of vines and flowers; it stands as a living tree, and he looks down at his new robe. It is purple and white, a robe fit for a King, and his feet

133

are dressed in shoes made of leather and silk. And then he throws his robe from his shoulders.

He is standing; his body is white and glistening with oils; he wears trunks made of silk, pure white, and girthing his waist is a belt made of silver, inset with diamonds and pearls; it is three inches wide, and the buckle of the belt is gold. There are wings, formed of gold leaf, attached to the heels of his shoes.

And then he takes onto his body whatever whimsy he feels: a fine spear appears in his hand like a staff, cartridge belts criss-cross his muscular chest, a spiked metal shield appears, strapped to his wrist; he holds it before his body. And a metal helmet covers his head, the visor is raised. A horse whinnies in the distance. And then he takes on studded leather wristbands, anklets of beautiful feathers; bright war paint appears on his chest: drawings of buffalo and dead warriors. His hips are girthed by a belt holding a scabbard; a broad sword sways at his side.

He looks then up at the cross; he raises his wrist and his shield before him; he takes his sword from its scabbard; he raises the sword and the shield in the air before him. And then he yells at the tops of his lungs; he casts the sword and the shield into the air, and he steps backward twenty feet, and drops into a crouch.

And then he rises into a slow run, his knees pumping high into the air. Bells and pieces of loose metal jingle in time to his stride, and when he reaches a point five feet from the cross he leaps up, his arms extended before him, and tangles his hands into the vines and flowers of the crosspiece of the cross, and begins to pull himself up. The muscles of his arms glisten with apparel and sweat, and he pulls himself up, until his head is above the vertical trunk of the cross, the crosspiece at a level with his groin, the space under his arms forming two triangles of air. And then he yells out: "Bring the nails!" And the nails appear behind him, suspended in the air, the three points of a triangle the size of his extended body, floating at a distance three feet from the cross.

And then he lowers his body for a moment slowly, then thrusts his arms straight again. And his body flies up straight, and he throws his arms over his head, spins in the air, and before he can fall he whips his arms out from his body, grabs hold of the vines of the crosspiece, and hangs from the cross facing the floating nails.

And then he throws his head; his helmet flies in the air; his hair falls around his shoulders. And then he looks up and smiles broadly at the empty sky. And when he lowers his head he sees the nails, and then he smiles again, yells: "Now!" And then he releases the branches and vines, opens his palms, and the nails come like bullets, ripping through the flesh of his hands and feet, fixing him to the wood of the cross.

And after a while he allows the people to move again; he bows his head under the black sky. He returns to himself, to the blood smeared, dying figure, upon the cross.

3 — The Jeweled Cross

The cross is carved in the hard ivory of a cameo; a small brooch, oval and convex, with the bust of a veiled woman carved upon it. The off-white color; the figure is his mother, the oval surrounded by tiny blue diamonds, set in yellowed silver.

The carved folds of the veil fall across her bosom, but the left side of her breast is slightly exposed. Her heart is visible and red; it is crowned with thorns; a small stiletto is driven through it, at an angle. And in the center of the heart, below the thorned crown, the drops of blood hanging from the thorns, is a small jeweled cross, imbedded in the heart's flesh. The ivory brooch floats on the conjunctive fluid covering the blue pupil of his right eye.

Only the whites of his eyes are visible; the eyes have rolled back in his death. The people are beginning to get restless. The soldiers are gathering their implements. He hangs limp on the cross. Both thieves are dead. A few of his followers are preparing to take him down.

His eyes look into his head. The left eye is dilated, glazed over, the eye's membrane is drying; the pupil covers the entire eye. But the pupil of the right eye is contracted, almost a pinpoint; his head is an empty vault; the inside of his skull is pure white; the lines of his skull's sutures are the only imperfections, and they are perfectly symmetrical and even. The walls of his white skull glow with an aura of blue light. His head is completely empty, a space. His brain rests, in light gray perfection, pillowed on a cloud, pulsing, somewhere in another sky.

The people are leaving. Only a few of his followers remain. A man stands on a barrel, beside the cross; he reaches up; he pulls on the Savior's eyelashes, closing his lids down over the backs of his eyes.

And now the inside of the skull glows brighter. There is no semblance of external light entering. The pupil of his right eye contracts even farther. The cameo floats, suspended, on the fluid on the pupil's center. And then the cameo moves, out from the pupil's surface, on a straight line; it floats in the blue-white light, cast in the empty vault of his skull. And now he can see it.

The off-white carving of the ivory cameo seems stamped and embossed into the hard bone of his skull. The oval is circled in a ring of yellowed silver. The deeper grooves in the carving of the veil seem yellowed also; the heart in the left of his mother's breast is wine red, crowned with thorns, woven in vines that appear like woven yellowed ivory. Blood sweats from the thorns; red drops hang from the blade of the shining stiletto, the handle of which is wrapped tightly in leather. And in the middle of the heart, below the crown of thorns, the Jeweled Cross winks out from its jewels and is imbedded in the heart's red flesh.

And he can see the cross clearly. It is fashioned from fine silver, hammered flat. And the four points of the cross have been carved and broadened, fashioned into silver clovers. Rubies, diamonds, moonstones, onyx, jade, opals, pearls, obsidian, coral are the jewels fixed to the cross, each in its own unique setting. The cross stands still in the pulsing heart of his mother.

And then in the pinpoint deep blue pupil of his eye, an infinitesimal figure takes shape, comes to his vision. He sees it is a human figure. It forms from the matter of his eye, the center of the pupil, as up from a deep blue pool, and floats on the viscous surface. It is a naked fetus. He sees the heart pulsing and weeping; blood drops fall from the stiletto, and the fetus moves out from the surface of his pupil; it floats in the still air of the vault of his glowing skull and begins to move. And the whole cameo seems to bend out from its silver frame; the figure of the fetus, tucked up in a ball, turns slowly across the void; it begins to
136

open, into a child; it is himself; his arms move in slow motion in the air; his legs kick in slow motion, as if he were running, though turning, slowly, graceful somersaults, moving even closer to the veiled figure, whose head is tilted slightly to one side. And as he moves he grows, and as his body takes on co-ordination it takes on clothing, various robes, sandals and capes, and begins to be able to right itself, growing though childhood, adolescence, and into a man.

There is an empty place in his pupil that undulates, changes, like a transforming piece of a jigsaw puzzle, and as the small body takes shape, moving in air, the void in his pupil takes shape also.

Now he has reached manhood; he stands, still in the air, halfway across the space that ends with the back of his skull, where the cameo sits, embossed in the hard white bone. But though he stands, arms at his sides, he moves still, but slowly, as if on some invisible conveyor belt, transversing the distance, between his eye and the ivory bust of his mother.

And then the figure in the air raises its arms, and the source of the figure, the pupil of the right eye, takes on the clearly defined shape of the figure with its arms raised. And then the figure turns slowly in the air of the vault directly before the Jeweled Cross inset in the red flesh of his mother's heart. And when the figure completes its turn, and is facing his eye, the clothing dissolves from the figure's body, and the body shows itself, ivory, haloed in blue light, hanging before the Jeweled Cross. And then the figure presses itself into the cross, and hangs, among blinking jewels, in the center of his mother's heart.

And the pupil of his right eye begins to contract; the blue pool floods into the void of the jigsaw space of the Jeweled Cross. At the back of the skull, the veil falls slowly over, obscuring the heart. The circle of blue diamonds in the yellowed silver setting fades away. The cameo bust of his mother fades, mixing its ivory into the white bone of his skull. The veil closes; his mother's face begins to dissolve. The blue light fades in the empty dome of his skull. The cornea is flooded with white ivory. The world, in his head, disappears.

PIETÀ

His body hangs before her, partially suspended in the air, his right hand closed now, in a fist, around the nail that holds his body to the cross, halfway erect: his left arm hanging loose across his body, dislocated, flapping lightly, like the edges of a drying garment on a line. His feet are also pinioned, to a block of wood, three feet above the ground.

She kneels, or genuflects before the cross, as they are lowering his body, emptied of its blood and growing cold to the touch. They slide it gently down the wood, to where she bends and kneels upon the ground, six feet away, and place him in her arms.

✝

Her body's form is bent before the cross, and from the back a simple oval form. Her hair is covered by the veil. The veil is blue and draped across her shoulders and her back. Her head is bowed, but as they slide him from the wood, her arms extend and reach out to receive his body slipping from their hands. One leg is bent behind his buttocks as he slowly falls, and then his body pitches forward, on the ground, before her kneeling figure, stirring up a rise of dust, that curls around his legs and arms and coats them lightly gray. His face is buried in the dust, two feet before her knees.

✝

One man stands upon a barrel, propped beside the cross, and wipes his face and chest and shoulders clean of blood and sweat. Two women wait below the cross, their arms extended, to receive his body as they take it down. They wait at either side, so that his mother can behold his total form. Another pulls his left

139

hand off the nail. The hand jerks loose too quickly, and the man upon the barrel accepts the arm that drapes around his shoulder; the half-hung corpse embraces him, and he begins to fall into the women's extended arms.

✝

He only wears a kind of loose-hung underwear, like swaddling clothes, white, but stained with blood and dirt. His skin is black and blue, his arms are dislocated at the elbows, wrists, and shoulders. The color of his skin is changing, as the blood drains out, into ashen gray. His mother's veil is blue; her garment, under it, is white. She stands, and bows three feet away. Her right hand reaches for her heart; she weeps into her left; she beats her breast, then sees the bowels through the open hole in his side, as they extend his stomach as they take him down to place him in her arms, and as she sees his bowels, she faints and falls upon the ground.

✝

She kneels upon the ground before the cross. They place his body in her arms. His legs are slightly spreading at the knees, his ankles dislocated, twisted in the dirt. His right arm dangles in the dust between her knees, the palm is open, facing upward from the ground. They place his body in her lap; his left arm swings and rolls out, and then becomes suspended from his shoulder in the air behind her back. The arm is dislocated at the shoulder, elbow, and the wrist: it traces out a single line of flesh behind her body. Then she reaches down to lift his right arm from the dirt, and as she bends to do it, he begins to slip, sliding from her knees; his flesh is sliding slowly through her fingers, and his body's weight, as she bends over him, begins to pull her forward, past the plumb line of her knees, and she begins to fall, upon him, on the ground.

✝

They lift him down, two women, and a man above them hang-
140

ing from the crossbeam of the cross, who fights to hold his wrist above his head, to guide the heavy corpse into their waiting arms. His wrists are gathered in the other's hands above his head; his head is bobbing back and forth, dislocated at the neck; the thorns are tearing at his biceps as he falls, begins to slip. His mother sits, four feet away, upon a barrel, weeping in her hands; and then extends her arms; her face turns up; the veil slips down from off her head and falls around her shoulders, and her hair unfolds and rolls down her back, and then spreads out in thick and raven-colored waves that swirl gently in the wind, and settle round her body, cover it in black, and trail in the dust.

✝

A woman wipes her face and then begins to make it up; she kneels in the dust, before his mother, who sits upon her heels before the cross. Another woman combs her raven hair and pins it back into a bun and pins the veil to her head. A man brings up a milking stool for her to sit on. Another man, whom she can't see, surrounded as she is, is wiping at the wound within the corpse's side, pushing back the bowels. And then he ties a velvet cord around the Savior's dislocated hips, and ties him firmly to the cross. And then the two men work upon his hands, and pull them off the nails. But his arms are dislocated at the shoulders, and his body falls, bending at his cinctured waist, until it hangs, his head straight down between his knees; his golden hair hangs down between his loose extended arms, stirring in the breeze, and brushes back and forth upon his pinioned feet. And then his mother rocks back on the stool, clutching at her breast, and they rush toward her as her body falls.

✝

One man hangs upon the crosspiece of the cross, working at the corpse's hand, pulling at the nail. The cross begins to vibrate from his diligence. He pulls and jerks.

She sees the shaking of his chest, the way the bowels are visible, churning slightly from the motion, in the concave chalice

141

of his hips. A woman is combing at her four-foot wealth of raven hair, which billows up, electric in the air. The veil gets
caught within the static field, begins to slip from off her shoulders, and she turns away to grasp it lest it strike the ground,
and doesn't see the cross is falling forward, with the body of her
son, extended, from his hands and feet; the bowing of his body
like a lovely masthead, or a sea bird, that hovers in the air a
moment, just before it drops its body, softly, on the bosom of a
swell. The man is tumbling in the air, a thick and heavy nail,
shining, in his hand.

✝

She sees the dead and lovely figure of her son from where she
sits, upon a milking stool, five feet before the cross. A blue veil
hangs down to her waist. Her face, devoid of obvious make-up,
dry of tears, looks upon the ashen body of her son. Her hands
are folded in her lap; she sits erect.

Her son is pinioned to the cross by hands and feet. His arms
are broken, dislocated, but extended in a lovely curving arc,
nailed, above the level of his head. His hips are flush against the
cross. His legs are slightly bent, the knees together. His left foot
rests upon his right, one nail holding both of them in place
upon the block. The crown rests perfect on his head, is centered,
and the blood has stopped. His mouth is slightly slack; his eyes
rolled back and creamy white, like oval cups of fluid, in his
perfect face.

✝

One man climbs the cross and sits upon the crossbeam, working
at the corpse's hand, and then removes the nail, and the body
slumps and seems to float, outward in the air, and then it comes
to rest against the body of a man who stands upon a barrel at
the cross's side; the loosened arm that falls, devoid of joints or
muscles, down his back, the armpit like a cool receptacle in
which his shoulder fits. His right arm gathers round the corpse's
waist; he holds him, comfortable and balanced, for a moment,

on his hip. And then the barrel begins to wobble, and his body
starts to slip.

✝

The stool sits firm upon the ground. She sits erect and opens
up her folded hands and rests them, with the palms up, on her
knees. The women stand away from her and hide their faces in
their veils. Her blue veil rests securely on her head; her head
is slightly tilted to the side and slightly lowered, looking at a
place before her breast. And then she takes her left hand from
her knee and opens up her garment and removes her breast.
The gown she wears beneath her veil is purple and of velvet,
and her breast protrudes, as if from a slightly opened curtain
on a stage. She puts her right hand back upon her knee, and
then she slowly turns her tilted head and looks upon the extended body of her dead, near naked son. Her arms rise slightly from her upper legs; her extended hands are open, reaching
slowly out. And from the cherry nipple of her breast, an oval
drop of cream comes forth, and hangs and slightly bobs, suspended in the air.

✝

Two women stand below the cross, their arms extended upward
to receive the corpse. A man climbs up the cross and starts to
work upon the nail that still remains. Another stands upon a
barrel, holds the corpse upon his shoulder and his hip. The
corpse's head rolls, slightly shaking from the action of the man
who's climbed the cross and now is working at the nail. The
barrel starts to wobble. Then the nail comes loose. The corpse's
arm collapses on the shoulder of the man who holds him, and
drapes across his back, and as the arm comes down, he turns
and lets the body slide into the women's waiting arms. The
corpse's feet are pinioned to a block of wood. The nail holds;
the corpse slips down, disjointed, from the women's hands. The
corpse's arms are dangling in the dust; the head is twisted at
the dislocated neck; the chest is flat upon the ground, the legs

143

are still in air, close together, nailed at the feet. She reaches for her purple garment, covering her breast.

✝

The man turns slightly, holding with his left hand to the cross. He lets the body slide from off his shoulder and his hip. His right hand slides along the corpse's back. The other man is working at the corpse's feet; the nail comes loose; a woman takes the feet.

✝

She sits alone, before the moving body of her son. Her head is slightly tilted to the side; her arms are now extended, slightly, in the air before her body. She sits erect, her veil hanging to her waist; her face is placid; tears run down her cheeks; her eyes are red; her veil begins to radiate, a light blue halo in the sun around her head.

✝

The corpse is turning in the air. A woman takes the feet. The man extends his arm, the body sliding, turning in its crook. The right arm of the corpse is dislocated at the shoulder, elbow, and the wrist; and as the body turns it falls, and circumscribes a single perfect arc, and then the upper body of the corpse slips down and comes to rest within a woman's arms; it rests between two women, slightly sinking at the waist, upon its back, suspended, slightly in the air.

✝

She sits erect upon the stool. And now they lift and place her son within her arms. His right hand falls across her chest; it comes to rest against the garment covering her breast. The left arm is extended, back behind her body, and the palm is facing upward; the lower arm is dislocated at the elbow and the wrist,

144

transcribes a perfect arc within the air. And now a halo forms around the corpse's head, beyond the thorns, but at a perfect distance from the crown, and where the head rests, like a child's, held within her arms, the back side of the halo lights her garment and her flesh. She turns his body slightly in her arms; their halos come together as she bends; they intersect each other, form a double-circled unit in the air.

And then she starts to rock him in her arms. And then the women turn and come to her. And then she starts to wail, and weep.

RESURRECTION

He comes to himself in mid-pace, discovers he is strolling. His right foot extends out before him. The nails reach only to the ends of the toes; they seem never to have been pared, but follow the curve of each toe. He notices, for the first time, that his second toe is longer than the large one. And he sees that his foot is naked, that his ankle is sinuous and beautiful; no veins show through its whiteness, like the virgin foot of a baby that has never come in contact with the ground. And then his toe points downward, as he takes his step, and sinks a few inches into some soft white substance, like cotton. His eyes move up the length of his body; he discovers he is completely naked, that his body is devoid of any mark. He has no navel; his sex is ambiguous; he feels his hair flow like silk over his shoulders as he raises his head.

And then he sees that he is walking along a cloud bank, a wide corridor, and that there is a continuous archway of clouds, over a mile above him, thick and billowy, like soft cotton; the clouds sink slightly under his feet as he strolls along. The place is bright, but the light has no color, and it casts no shadow of his body; there is no evidence of its source.

And then, as he looks down the corridor before him, he sees an object resting on a fluffy pillow of clouds to his left, and as he comes closer to it, he discovers it is a brain. And when he reaches the place where the brain rests on a soft-edged pillar of clouds, he stops and begins to look at it. It is a human brain; it pulses; it is light gray in color. And he begins to notice that there are tiny objects moving, over the brain's surface into its deep grooves, appearing and disappearing, into the brain's core. And so he picks up the brain, and holds it in the palm of his hand; it is heavy and dry.

He stands in the soft clouds, holding the brain in his hand, and as he looks closer he sees that the objects on the brain are

147

people, machines, and animals, moving across the cortex, into its folds and grooves, teeming over the surface of the brain. And he recognizes familiar events.

In one groove he sees the tiny figure of his father, sawing at a board, then fixing the board to another with pegs. He sees his apostles, himself washing their feet; he sees a familiar woman, rummaging in a chest. He sees himself: at a wedding feast, driving through a desert in a mechanical conveyance, touching his fingers to the wounds of people, all the various events of his life.

And then he discovers there is a pattern of activity on the brain's surface, and that all the objects remain there only a measured time, and that they always return to the brain's center. And so he grasps the brain, like a piece of fruit in his hands; he presses his thumbs into the brain and opens it. Inside the brain is an open space, a core; and in the core he finds all the figures he'd seen on the surface. The machines are all broken; the animals founder with broken and diseased limbs; the people move with the aid of crutches; some rest in beds; others move about in wheelchairs. And he sees they are all weeping. And then he looks closely, gives careful study to each infirmity, and he feels a particular sorrow, specific to each. And then he realizes, that though the sorrow really touches him, he remains whole and complete, even as he empathizes in various particular ways. He closes the brain then, with a smile, and places it back on its pillar of clouds. And then he begins strolling again.

And after he has walked for a while, he becomes aware of a possible source of the light around him. Before him, a good two miles away over the rolling clouds, he can see that the light is brighter, and that the corridor turns. He kicks at the fluffy clouds as he strolls, dislodging small puffs, like cotton, that arise and float in the air and then bounce down to their source, and settle. And when he has kicked one piece of cloud higher than the others, he hears a sound behind him and turns around.

At first he can see nothing, but then he sees a small pin-point of motion, a patch of brown, growing larger, coming toward him, over the fluffy clouds at a great distance. And then the

148

patch takes on shape, becomes furry. And then he drops to his knees, sinks in the clouds, raises his arms, and whistles through his teeth.

And Hound bounds down the corridor, leaping across the clouds, tumbling sometimes then rising and racing toward his Master. And when he comes within five feet of Jesus, he stops, extends his right paw into the cloud before him and lowers his head in a bow. And Jesus claps his hands together, and Hound leaps over the remaining distance, landing in the Savior's arms. And Jesus scratches his ears and strokes his neck and speaks to him quietly.

And after they have made their welcome, the two set out to complete their journey. Jesus strolls on the puffy clouds; Hound frisks at his heels. Occasionally the Savior kicks at a puff of cloud, and Hound chases it, and it comes to rest again, mingling with the clouds at their feet. The dog is naked, and no longer wears his cape.

And now they have reached the place where the corridor turns. There is nothing but a wall of soft clouds before them. The dog walks now, more alertly; he prances slightly, his head and tail erect, at his Master's side; his feet move like in stately dance, in the soft clouds. And then they turn the corner and come to face the source of light.

There are two gold gates before them; and Jesus looks up and sees that he cannot see the tops of the gates, that the vault of the corridor disappears in the sky now, that the tops of the gates are lost in the clouds. The gates are massive, but they are not heavy; they are like garden gates: intricate, woven structures of thin gold tubular metal; like a huge, delicate headboard in a sunlit feminine bedroom. And the intricate woven metal is intertwined with vines and flowers. But the vines and flowers are in no way natural; they are made in proportion to the massive size of the gates. Very large metal butterflies and insects flit round the gates; large bees suck at the nectar of flowers. And a light emanates from the gates. And Jesus and Hound are dwarfed by them.

And then the gates, slowly, begin to open, making no sound at all; they swing about five feet open, allowing enough room for a man and a prancing dog to pass. And then Jesus and Hound see that beyond the gates is a meadow, and that the meadow falls away from the gates and spreads out almost as far as their eyes can see. The meadow is growing with low grasses and flowers; is a high altitude meadow, above timber line; and at a great distance it is surrounded by mountains of rock.

And in the center of the meadow, about one hundred yards away, there is a huge walnut tree, and under the tree is a kind of throne, like an easy chair. And Jesus sees that his real father is sitting in it.

His father is large and old, but he sits straight up in the throne. The throne has large gilded wheels at its sides, and when his father sees him, he waves at him, then reaches down, and begins to wheel the throne toward his son.

The father moves in his throne through the meadow. The son strolls toward his father, waving. The dog prances and barks by the son's side.

And when he comes face to face with his father, he bends and falls into his father's arms. The father is naked also, but his lap is covered with a beautiful veil. In the center of the veil is a half-carved walnut, surrounded by parings. Some delicate carpenter's tools rest in the veil also. The outer shell of the nut has been carved with intricate lines and concavities; it looks like a miniature globe of the earth, like a small brain, like a round map or legend at a lookout point. As the son's face lowers into the father's lap the father removes his tools and the half-carved walnut, placing them in a basket affixed to the side of the throne. And then the son's face enters his lap. With his right hand he strokes the head of his son; with his left hand he scratches the ears of the dog, who now stands quietly at the foot of the throne.

And after a while the father speaks. "Now look," he says — and Jesus raises his head to look at him — "your mother will be here in a while. But before she comes, let's go down there into the meadow for a while, under that good shade tree. I have something to show you. And we can talk."

And the son wheels his father down to the meadow, to the place where the walnut tree offers shade. And below the tree is
150

a small rectangular pool, a hot springs bath, about three feet wide, the sides and the bottom cemented with moss-covered stone. The grasses and flowers come up to the edge of the spring. The pool is five feet deep, and on either end of the pool, two feet under the hot clear water, a large thick root from the walnut tree passes through the pool; the roots are stripped of their bark, are like polished ivory, are good seats.

And then his father smiles over his shoulder at Jesus, and directs him to wheel the throne over to the side of the spring. And then Jesus offers his back to his father, who grasps his son round the neck. And Jesus lifts the light weight of his father's body. "Now back me over the edge of the spring," his father directs him. And Jesus does this. "Now let me fall," and Jesus drops his father, who lands with a splash in the soft water, laughing.

And then Jesus turns round, and he sees that his father is submerged in the water of the spring up to his neck, that he is comfortably seated on a root at one end, that his two arms rest in the grasses and flowers on the pool's sides. "Come on," his father says, "bring Hound with you." And Jesus lifts up Hound in his arms; he extends his leg; his ankle seems to dislocate; his foot extends down toward the water's surface; he touches the top of the water with his longest toe. And then he steps in with the dog in his arms. He places his foot on the root; he lowers himself into the rich clear soft warmth of the spring.

He sits on the root; Hound rests in his arms; the water washes over their shoulders. The father sits facing the son and his dog.

And then they begin to talk to each other. His father takes up his walnut and a small delicate knife and begins to carve, his elbows resting on the sides of the pool; occasionally, he dips the nut under the water's surface; the fragments of carved flesh from the shell float between them.

And after they have spoken of the clouds, of the great gates, of the pleasure of the tree and the pool, Jesus begins to stroke Hound's head; and then his father stops carving and looks at him. He reaches out, brushing the bits of shell to the pool's side, clearing the water between them. And then Jesus begins to speak to his father. And his father sits back and becomes still; his elbows rest on the sides of the pool. Occasional walnuts, fall with faint thuds, in the grasses around them. And his father listens to him, attentively, as Jesus tells him, the story of his Journey, in another world.